DATE DUE

About Island Press

Island Press is the only nonprofit organization in the United States whose principal purpose is the publication of books on environmental issues and natural resource management. We provide solutions-oriented information to professionals, public officials, business and community leaders, and concerned citizens who are shaping responses to environmental problems.

In 1994, Island Press celebrated its tenth anniversary as the leading provider of timely and practical books that take a multidisciplinary approach to critical environmental concerns. Our growing list of titles reflects our commitment to bringing the best of an expanding body of literature to the environmental community throughout North America and the world.

Support for Island Press is provided by The Geraldine R. Dodge Foundation, The Energy Foundation, The Ford Foundation, William and Flora Hewlett Foundation, The James Irvine Foundation, The John D. and Catherine T. MacArthur Foundation, The Andrew W. Mellon Foundation, The Pew Charitable Trusts, The Rockefeller Brothers Fund, The Tides Foundation, Turner Foundation, Inc., The Rockefeller Philanthropic Collaborative, Inc., and individual donors.

About IUCN—The World Conservation Union

Founded in 1948, The World Conservation Union brings together states, government agencies, and a diverse range of nongovernmental organizations in a unique world partnership: more than 800 members in all, spread across some 125 countries.

As a union, IUCN seeks to influence, encourage, and assist societies throughout the world to conserve the integrity and diversity of nature and to ensure that any use of natural resources is equitable and ecologically sustainable. A central secretariat coordinates the IUCN Program and serves the union membership, representing their views on the world stage and providing them with the strategies, services, scientific knowledge, and technical support they need to achieve their goals. Through its six commissions, IUCN draws together more than 6000 expert volunteers in project teams and action groups, focusing in particular on species and biodiversity conservation and the management of habitats and natural resources. The union has helped many countries to prepare national conservation strategies and demonstrates the application of its knowledge through the field projects it supervises. Operations are increasingly decentralized and are carried forward by an expanding network of regional and country offices, located principally in developing countries.

The World Conservation Union builds on the strengths of its members, networks, and partners to enhance its capacity and to support global alliance to safeguard natural resources at local, regional, and global levels.

EXPANDING
PARTNERSHIPS
IN
CONSERVATION

EXPANDING PARTNERSHIPS
IN
CONSERVATION

Edited by Jeffrey A. McNeely

IUCN—The World Conservation Union

ISLAND PRESS
Washington, D.C. ❑ Covelo, California

The views and opinions expressed in this work do not necessarily represent those
of the Commission of the European Communities, IUCN, or other participating
organizations.

Published by: Island Press, 1718 Connecticut Avenue, N.W., Suite 300, Washington, DC 20009.

ISLAND PRESS is a trademark of The Center for Resource Economics.

Library of Congress Cataloging-in-Publication Data

Expanding partnerships in conservation / edited by Jeffrey A. McNeely
   (International Union for the Conservation of Nature and Natural
   Resources, IUCN).
       p.   cm.
    Revised papers from the IVth World Congress on National Parks and
   Protected Areas, held February 10–21, 1992, in Caracas, Venezuela.
   Includes bibliographical references and index.
   ISBN 1-55963-351-4 (paper)
    1. Nature conservation—Planning—Congresses. 2. Natural areas—
   Planning—Congresses. 3. Nature conservation—Citizen
   participation—Congresses. I. McNeely, Jeffrey A.
   II. International Union for Conservation of Nature and Natural
   Resources. III. World Congress on National Parks and Protected
   Areas (4th : 1992 : Caracas, Venezuela)
   QH75.A1E95   1995
   333.7'2—dc20   95-19740
                                                             CIP

Printed on recycled, acid-free paper ⊛

Manufactured in the United States of America
10 9 8 7 6 5 4 3 2 1

Contents

Part II Partnerships with Major Sectors

Part III Partnerships with Communities

Contributors

OLAYIWOLA AKERELE
World Health Organization
Traditional Medicine
CH-1211 Geneva 27
Switzerland

J. L. ANDERSON
Kangwane Parks Corporation
PO Box 1990, Nespruit 1200
South Africa

BETI ASTOLFI
UNIFEM
304 East 45th Street
New York, NY 10017
United States

JAMES R. BARBORAK
Wildlands Management
NYZS / Wildlife Conservation
Society, University for Peace
Apartado 277, Heredia 3000
Costa Rica

DEVIN M. BARTLEY
Inland Water Resources and
Aquaculture Service, FAO
Viale Delle Terme di Caracalla
00100 Rome
Italy

STEPHEN O. BENDER
Organization of American States
1889 F Street, NW
Room GSB 340 I
Washington, DC 20006
United States

LAURA CAMPOSBASSO
WWF-US
1250 24th Street, NW
Washington, DC 20037
United States

MICHAEL COHEN
Eastern Cape Region
Private Bag X1126
Port Elizabeth 6000
South Africa

KENNETH COX
Canadian Wetlands Conservation
1750 Courtwood Crescent
Suite 200
Ottawa, Ontario K2C 2B5
Canada

MICHAEL DOWER
Peak National Park
Aldern House, Baslow Road
Bakewell, Derbyshire DE4 1AE
United Kingdom

EUSTACE D'SOUZA
WWF-India
PO Box 3058, 172 B Lodi Road
New Delhi 110 003
India

MOHAMED T. EL-ASHRY
The World Bank
1818 H Street, NW
Washington, DC 20036
United States

B.C.Y. FREEZAILAH
International Tropical Timber
5F International Organization
Centre Pacifico-Yokohama
1-1 Minato Mirai, Nishi-ku
Yokohama 220
Japan

CHANDRA P. GURUNG
Annapurna Conservation
King Mahendra Trust for Nature
Jawalakhel, PO Box 3712
Kathmandu, Nepal

PARVEZ HASSAN
Paaf Building
7D Kashmir Egerton Road
Lahore, Pakistan

JAMES M. KAPETSKY
Inland Water Resources and
Aquaculture Service, FAO
Viale Delle Terme di Caracalla
00100 Rome
Italy

ANNETTE LEES
Maruia Society Inc.
143 Bethells Road, RO1
Henderson, Auckland 8
New Zealand

WALTER J. LUSIGI
Technical Department, African
Region, The World Bank
1818 H Street, NW
Washington, DC 20036
United States

GARY E. MACHLIS
Department of Forest Resources
College of Forestry
University of Idaho
Moscow, ID 83843
United States

EDWARD MALTBY
Wetland Ecosystems Research
Group, Royal Holloway Institute
for Environmental Research
University of London, Huntersdale
Callow Hill, Virginia Water
Surrey GU25 4LN
England

SIMON C. METCALFE
Cornell University
No. 1, 410 W. Green Street
Ithaca, NY 14850
United States

DAVID A. MUNRO
2513 Amherst Avenue
Sidney, B.C. V8L 2H3
Canada

WILL MURRAY
The Nature Conservancy
2060 Broadway, Suite 230
Boulder, CO 80302
United States

TAKUYA NAGAMI
Environment Agency
1-2-2 Kasumigaseki, Chiyoda-ku
Tokyo
Japan

RUTH NORRIS
1310 S. Carolina Ave., SE
Washington, DC 20003
United States

GEORGE B. RABB
SSC, Chicago
Zoological Society
Brookfield Zoo
3300 Golf Road
Brookfield, IL 60513
United States

JOHN SCHELHAS*
School of Renewable
Natural Resources
University of Arizona
Tucson, AZ 85721
USA

RONALD G. SEALE
Government of the
Northwest Territories
Yellowknife, N.W.T. K1A 2L9
Canada

STANLEY SELENGUT
Maho Bay Camps
17 E. 73rd Street
New York, NY 10021
United States

WILLIAM W. SHAW
School of Renewable
Natural Resources
University of Arizona
Tucson, AZ 85721
United States

DEBORAH SNELSON
African Wildlife Foundation
PO Box 48177
Nairobi, Kenya

W. J. SYRATT
BP Engineering, Uxbridge One
1 Harefield Road
Uxbridge, Middx. UB8 1PD
United Kingdom

BERND VON DROSTE
World Heritage Centre
UNESCO
1 rue Miollis
75015 Paris
France

ERVIN H. ZUBE
325 BioScience East
University of Arizona
Tucson, AZ 85721
United States

*John Schelhas is currently at the Department of Natural Resources,
Cornell University

Preface

A World Congress on National Parks and Protected Areas has been held each decade since 1962. The objective of the congress process is to promote the development and most effective management of the world's natural habitats so that they can make optimal contributions to sustaining human society. The IVth Congress, held in Caracas, Venezuela February 10–21, 1992, aimed to reach out to and to influence numerous other sectors-beyond those professionals directly concerned with protected areas: Management agencies, nongovernmental conservation organizations, traditional peoples' groups, relevant industries, and resource managers were brought together and involved to enhance the role of protected areas in sustaining society, under the theme "Parks for Life."

The subject of partnerships was seen as so important that the topic was made the subject of a major plenary session, during which most of these papers were presented. Many of the topics were subsequently discussed at the various workshops held during the congress, and the papers were then substantially revised by the respective authors. The ideas put forward in these pages provide a distillation of the partnerships that can be formed in the search for sustainable relationships between people and resources.

"Cooperation" and "partnership" are very popular words these days, as the benefits of working together become increasingly apparent in a time of scarce resources for conservation. This book is meant as a contribution toward indicating the kinds of partnerships that can help support protected areas.

The convening of the IVth Congress was built on the very strong support of the government of Venezuela. Enrique Colmenares Finol, Venezuela's Minister for the Environment, served as co-chair of the congress and was a most gracious host to the participants. Local arrangements were efficiently handled by the Venezuelan Organizing Committee, under the able chairmanship of José Joaquín Cabrero Malo. Cristina Pardo, CNPPA's Regional Vice-Chair for South America, deserves a special vote of thanks for her steadfast support throughout the preparations for the congress.

A large number of partners—ranging from governments to private foundations—provided the financial resources necessary to organize and hold the congress. Bilateral assistance came from the governments of Venezuela, the Netherlands, Sweden (SIDA), Finland (FINNIDA), Germany (BMZ-GTZ), Norway (Ministry of Environment), Denmark (DANIDA), the United States of America (U.S. Department of State and Department of Interior, National Park Service), the United Kingdom (ODA), Switzerland (DDA and Intercoopération), Canada (Canadian Park Service), and France (Ministries of Foreign Affairs, Cooperation, and the Environment). Multilateral institutions contributing included the Commission of the European Communities (CE); Inter-American Development Bank (IDB); The World Bank; United Nations Educational, Scientific, and Cultural Organization (UNESCO); the World Heritage Committee; United Nations Environment Program (UNEP); Food and Agricultural Organization of the United Nations (FAO); United Nations Development Program (UNDP); and Agence de Coopération Culturelle et Technique (ACCT). International nongovernmental organizations and foundations supporting the congress include The Nature Conservancy, World Wide Fund for Nature (WWF), the MacArthur Foundation, and the World Conservation Monitoring Center (WCMC). British Petroleum helped support congress documentation. The World Resources Institute and the Bureau of the Ramsar Convention provided important services to the congress. Other institutions too numerous to mention provided support in kind; all are thanked for their contributions.

The manuscript production of this volume owes much to Sue Rallo, who worked tirelessly on the innumerable drafts; Morag White, who freely provided her considerable professional skills; and Caroline Martinet, who helped hold all the pieces together. This work was made possible in part by a grant from the Commission of European Communities.

Jeffrey A. McNeely
Secretary General
IVth World Congress on National Parks and Protected Areas
Gland, Switzerland

Chapter 1

Partnerships for Conservation: An Introduction

Jeffrey A. McNeely

Introduction

These are trying times for our planet, as the combination of growing human populations and increasing consumption (especially in the wealthy countries) is overwhelming efforts to conserve the biological systems on which all life depends. Obvious manifestations of the problems—loss of forests, species extinctions, water pollution, oil and chemical spills, acid precipitation, ozone depletion, uncontrolled urbanization, and destructive civil strife—evoke greater public concern; in places, a greater political will to implement the action is required to enable people to live in balance with resources. Conservation action at the international level can be expected to grow following the IVth World Congress on National Parks and Protected Areas (held in Caracas, Venezuela, in February 1992), the United Nations Conference on Environment and Development (held in Rio de Janeiro, Brazil, in June 1992), and numerous other consensus-building exercises in all parts of the world. The new Convention on Biological Diversity, the Global Environment Facility, and greater investments in conservation on the part of multilateral and bilateral development agencies are all positive indicators of expanding support for conservation.

Yet all of this is still far short of what is required to bring about sustainable relationships between people and resources on a global level. A crucial foundation of future action is people and institutions working together. Drawing on papers presented at the Caracas congress, this book describes how new and stronger partnerships can be formed, particularly between protected area managers and other sectors of society. It does not pretend to be exhaustive; the potential for productive partnerships is virtually unlimited. It does, however, indicate the kinds of activities that currently are being undertaken in

1

many parts of the world to build stronger partnerships to support conservation at international, national, and local levels.

Protected Areas: Contributions and Challenges

As development has accelerated in the past several decades, governments have come to recognize that legally protected areas can play an important role in the overall pattern of national land use and economic development. Their specific contributions to the well-being of societies include:

- Maintaining those essential ecological processes that depend on natural ecosystems
- Preserving the diversity of species and the genetic variation within them
- Maintaining the productive capacities of ecosystems
- Preserving historic and cultural features of importance to the traditional lifestyles and well-being of local peoples
- Safeguarding habitats critical for the sustainable use of species
- Securing landscapes and wildlife that enrich human experience through their beauty
- Providing opportunities for community development, scientific research, education, training, recreation, tourism, and mitigation of the forces of natural hazards
- Serving as sources of national pride and human inspiration

The value of protected areas may be even greater in the future. Preserving genetic raw materials will sustain future biotechnological advances in the fields of medicine, agriculture, and forestry. Protecting extensive naturally functioning ecosystems will be vital for monitoring global change and guiding human adaptation to a changing world. Protected areas are a major means of implementing the Convention on Biological Diversity; they will contribute to the sustainable forms of forestry envisaged in discussions about a possible new forest convention. Protected areas will also help maintain options for unforeseen future uses, thereby providing an "insurance function."

Despite the many important contributions protected areas make to modern societies, they suffer from a number of problems. While these vary from country to country, the most important general problems include:

- *Weak national constituency.* The numerous benefits of protected areas are seldom fully appreciated by the general public, because such areas are seen as exotic vacation spots or remote wildernesses rather than as essential contributors to national welfare. The lack of a strong national and local constituency translates into insufficient human and financial resources being devoted to protected area management. The other major problems faced by protected areas, described below, can be traced back to this fundamental issue of inadequate national and local support.

• *Conflicts with local people.* Establishing a protected area often requires explicit restrictions in the use of the area's resources by local people in the interests of the nation and future generations (hence the use of terms such as "national park" and "world heritage"). Insufficient attention has been given to enabling the local people to earn appropriate benefits from conservation programs, a problem made worse by the growing human populations in many rural areas. The balance between conservation for long-term national public benefit and exploitation for immediate local or private gain is elusive and constantly shifting. Conflicts often result if the local people lose personal opportunities when new conservation regulations are imposed on them without alternatives being provided to meet their basic needs. In the most extreme cases, local people can be actively hostile to the protected area, leading to vandalism and loss of life in pitched battles between "encroachers" and protected area staff. On the other hand, productive partnerships can be formed with local landowners and communities as they become more aware of the contribution their area makes to the national heritage, and as governments support forms of management that can provide benefits to both the people living in the area (including landowners) and the nation at large.

• *Conflicts with other government agencies.* The agencies responsible for protected areas tend to be relatively weak in the government structure, leaving them vulnerable to policy conflicts and budget cuts. Adequate legislative support is often lacking. Agricultural, forestry, and fisheries incentives may promote encroachment on protected areas (though they can also be used to help protect them); highway departments may find the "free" lands contained in protected areas attractive for new roads; tourism departments may try to attract more tourists than a protected area can support without damage to the resource; irrigation and energy departments may wish to build dams or drill for petroleum in protected areas, which often are "unoccupied" public lands and hence easy to use for these purposes; mining interests may wish to exploit mineral resources found in protected areas, for the same reason; and industrial development policies may stimulate pollution and associated climate change that adversely affect protected areas. Such policies as frontier settlement programs, planned colonization of protected areas for national security reasons, and commercial exploitation of natural resources to service national debts result from government decisions that seem to be oblivious of protected area objectives. But conflicts within government are not necessarily inevitable. Dialogue between the concerned sectors can often turn conflict into cooperation.

• *Insufficient management.* Protected areas are sometimes surrounded by agricultural lands or heavily fished areas and reduced to fragments of formerly extensive areas of habitat. In such cases the protected area manager must take an active role in managing the remaining habitats, the species they support, and the way people use species and habitats. Small populations of wildlife may not be viable in the long term without such management. Possible changes in climate and resource use in surrounding areas add to the management challenge. The scientific basis for effective management is often lacking, and the required trained manpower is not always available to implement even simple management measures, let alone the sophisticated interventions called for by modern pressures on species and habitats. In the past, many managers have also considered their challenges to be primarily ecological rather than social, economic, and political; they have thus considered their management problems in a narrow ecological sense rather than in terms involving adjacent areas, local people, and other sectors. This problem, too, calls for forming partnerships with the broader community and the specific sectors most closely affected by protected area management.

• *Insecure and insufficient funding.* Most protected areas are funded from the national budget. Many demands are made on national funds, and protected areas are often a poor relation receiving declining shares of the budget. Most countries find it difficult to justify increased expenditures on protected area management, which may be accompanied by higher indirect costs at a local level, and by still higher local and regional opportunity costs. The link with economic development is seen as too remote, the diversion of other program funds is seen as too expensive in the short term, and the potential land use conflicts with local government and local populations are seen as too troublesome. Perhaps worse, even where protected areas are highly profitable (tourism to Kenya's national parks, for example, is that nation's second leading foreign exchange earner), only a small portion of the economic benefit they earn is reinvested in the management of the protected areas or in the welfare of communities in the surrounding lands. But if, as this book argues, many other sectors are earning benefits from protected areas, it would seem that considerable scope exists for broadening the base of financial support for these areas. This is especially likely to be the case when a closer link can be formed with development objectives, and where local people support the protected area.

Meeting the Challenges: Ten Principles
for Successful Partnerships

In responding to these problems, many governments, nongovernmental organizations, and businesses are seeking innovative ways to add new protected areas, improve the management of the existing areas, and build more positive relationships with the people who live in and around the protected areas—all of this in a time when available budgets are shrinking and demands on resources are increasing. These new approaches to establishing partnerships for improving the management of protected areas build on the following 10 principles:

1. *Provide benefits to local people*. Popular and political support for a system of protected areas is strengthened when it generates a flow of public benefits to people. The more people benefit directly from the protected areas, the greater the incentive for them to protect the resource and the lower the cost to government of doing so. If a government decides to establish a site as a protected area and this would cost lost opportunities for local communities, then these local communities should be compensated for their losses. The benefit–cost ratio of conserving a protected area must ultimately be positive for the local people if the area is to prosper in the long term, and this will require that the local people be appropriately involved in the planning and management of the protected areas and that they share in the benefits. Precisely how this is to occur will vary considerably from place to place. Some approaches that have been effective are suggested in Chapters 10 (Akerele), 14 (Bender), 26 (Gurung), 27 (Seale), 30 (Anderson), and 32 (Snelson). Chapter 17 (El-Ashry) makes the point that development agencies should include the provision of benefits to local people as an essential element of their support to protected areas.

2. *Meet local needs*. Legislation, management policy, and operational practice for protected areas must facilitate both the satisfaction of local needs and the wider goal of conserving biodiversity, as described in Chapter 6 (Hassan). For example, the legislation prohibiting collection of material for study, which applies in some national parks, actually hampers the evaluation of their biological diversity; and some regulations forbid the interventions required to provide sustainable benefits to local people or to manage certain species. Such restrictions may need to be reviewed and revised. Measures to provide local benefits, and to enable translocation of animals and plants for protection and restoration of species and habitats, need to be provided in protected areas in forms that are appropriate to the objectives of each individual site.

3. *Plan holistically.* Management of a protected area and that of adjacent areas must be planned together. Few protected areas can be self-contained, isolated entities, and some are managed landscapes where people live and work. The integrity of a strictly protected core zone is often dependent on "transition zones" in which human uses of natural resources around protected areas are compatible with conservation objectives. Transition zones must be managed so as to harmonize social goals in the core of protected areas and in the surrounding lands, as described in Chapters 3 (Lusigi), 8 (von Droste), 20 (Zube), and 31 (Metcalfe). Such transition zones may be managed by forestry (Chapter 9, Freezailah), private landowners (Chapters 28, Cox; 29, Cohen; and 31, Metcalfe), water resources agencies (Chapter 13, Dugan and Maltby), or even the military (Chapter 19, D'Souza).

4. *Plan protected areas as a system.* Protected areas need to be conceived and managed as a system that addresses national and international objectives for meeting the needs of sustainable societies. Such a system needs to include appropriate management of privately owned lands that are important for conservation, through means such as those described in Chapters 22 (Lees), 23 (Murray), 24 (Schelhas and Shaw), 28 (Cox), and 29 (Cohen). As suggested above, all of a protected area's neighbors need to be considered in the system plan.

5. *Define objectives for management.* Many different administrative approaches can be taken to managing land and sea to conserve nature while contributing to sustainable development, depending on the desired management objectives. Under IUCN's system of protected area categories, Category II National Parks by definition need to be protected against extractive resource exploitation on a commercial scale, while Habitat/Species Management Areas (Category IV) and Protected Landscapes/Seascapes (Category V) can be used more flexibly to permit a range of human activities that are consistent with the conservation objectives of the area. Such areas can protect traditional forms of agriculture or fishing (which often contain high biodiversity), as an integral part of a nation's protected area system. The importance of this role is shown in Chapters 3 (Lusigi), 12 (Kapetsky and Bartley), 26 (Gurung), and 31 (Metcalfe).

6. *Plan site management individually, with linkages to the system.* Each protected area is different in terms of species and habitats, local human populations, history, climate, and a range of other factors. Therefore, its management needs to be site-specific and the local staff, communities, and interest groups should be involved in planning that management. Perhaps even more important, the knowledge that local people have about the area and its

resources should be utilized in management programs to meet local, national, and international needs. Ways of building on local knowledge are described in Chapters 3 (Lusigi), 10 (Akerele), 18 (Astolfi), 22 (Lees), 25 (Dower), and 31 (Metcalfe).

7.   *Manage adaptively.* Conditions are changing quickly, so each management plan for an area needs to incorporate the capacity to adapt to changing conditions—climate, economic conditions, population, developments in surrounding areas, war and civil strife, and so forth. A dynamic planning process is required, involving wide consultation with interested parties rather than the production of rigid plans. Ways of involving partners in implementing management plans are addressed in Chapters 2 (Munro), 3 (Lusigi), 5 (Barborak), and 7 (Machlis).

8.   *Foster scientific research.* Research in both the natural and the social sciences is needed by protected area managers to assess basic ecological relationships, dynamics of change, needs of local people, possible results of habitat manipulation, effects of tourism, and so forth. Further, managers must consider all ecosystem management procedures as scientific experiments to be monitored continuously as the effects of such procedures become apparent. Chapter 7, by Machlis, provides advice on this subject, as do several of the other publications arising from the Caracas Congress, notably *Research in Protected Areas*, edited by David Harmon and published as a special issue of the *George Wright Forum* (9: 3–4, 1992).

9.   *Form networks of supporting institutions.* Central governments alone cannot carry the full responsibility for conserving nature, and a range of different institutional arrangements can contribute to national conservation goals. A complex and diverse array of institutional arrangements is required to manage protected areas for meeting society's needs. As described in Chapter 5, by Barborak, these will include national, regional, and local government agencies; universities; private landowners; NGOs; private businesses; cooperatives for common lands, freshwater, and coastal areas; and others. Chapters 21 (Norris and Camposbasso), 22 (Lees), and 23 (Murray) provide examples of the roles nongovernmental organizations (NGOs) can play in such comprehensive approaches.

10.   *Build public support.* Public support for protected areas is essential. Great efforts are required from numerous sectors to ensure that information about how protected areas are meeting society's needs is communicated through the mass media, and that universities, museums, zoos, aquaria, and botanic gardens are given opportunities to help communicate this message. Chapters 2 (Munro), 3 (Lusigi), and 11 (Rabb) highlight methods and examples to achieve this greater support.

Converting Principles into Practice:
The Caracas Action Plan

This book seeks to elaborate on the 10 principles for participation, showing how they have been implemented in practice and how partnerships can be formed better in the future. Further guidance is provided by the Caracas Action Plan, produced as a result of the Parks Congress and published in *Parks for Life* (IUCN, 1993). One of the four main elements in the Caracas Action Plan is building partnerships for conservation. That element recommended three main procedures that protected area institutions can follow for building partnerships.

First, *identify the key protected area interests of various groups.*

- Determine the potential range of products and services that can be provided by protected areas, including those relevant to nontraditional interest groups such as religious groups, artisans, users of traditional medicines, and the military.

- Identify the groups that have a stake in these services and products. Promote the sharing of views and experiences, and the development of organizations to represent the interests of groups not yet organized.

- Explore means for enhancing the benefits obtained by some groups from protected areas without diminishing those of others.

Second, *recognize priority concerns for local communities.*

- Work with local communities to determine how management of the protected area can help meet local needs. Develop an understanding of local resource issues through building on local knowledge and perceptions of needs. Develop consultative processes that encourage competing groups to identity optimal management solutions acceptable to a majority.

- Promote attitudes among protected area managers that encourage recognition of the need of local communities for equitable and sustainable development.

- Seek the support of local communities in promoting protected areas by offering opportunities for influencing decision making—for example, through representation on local protected area management boards and at public debates on management issues.

- Based on examples of success, publish guidelines for establishing co-management and co-financing arrangements that take into account all interested groups.

- Develop participatory research involving local people and institutions as a tool for planning, a means of sharing basic information, and a mechanism for building working relations among interest groups.

Third, *stimulate informed advocacy.*

- Assess the vested interests of groups and take account of these in seeking greater political and financial support for protected area programs.

- Reinforce the support of key interested groups through award schemes, public ceremonies, and personal communications.

- Strengthen education and information programs within protected areas, and widely disseminate information on protected area issues.

Conclusion

Protected areas will prosper only if they are supported by the public, the private sector, and the full range of government agencies. This support is most likely to be forthcoming when all parts of society are aware of the importance of protected areas to their own interests, when the protected areas are well managed and contribute to the welfare of the nation in a cost-effective way, and when the public is aware of the contributions that the protected areas are making to their lives and to the society in which they live.

On the other hand, the modern approach to protected area management, involving partnerships with local human communities, faces formidable challenges. Many protected area staff believe that the cooperative approach could ultimately reduce the quality of the protected area, and that strong legislation supported by vigorous law enforcement is the best option for long-term conservation. And indeed, experience has shown that local people often are as likely to misuse privileges under cooperative management as anyone else. Even so, given the insufficient staff and logistics support available to most protected areas, the "strict preservationist approach" is both impossible to implement and of doubtful validity on conservation grounds. The conciliatory and cooperative approach advocated in this book may be the only viable option in today's conditions.

We are at a crossroads in the history of human civilization. Our actions in the next few years will determine whether we take a road toward a chaotic future characterized by overexploitation and abuse of our biological resources, or take the opposite road toward maintaining biological diversity and using biological resources sustainably. The partnerships that are formed today, such as those described in this book, demonstrate that protected areas—in their many forms—can contribute to guiding people in their choice by providing many goods and services of long-term benefit to human society.

Part I

PRINCIPLES OF PARTNERSHIPS

Chapter 2

New Partners in Conservation: How to Expand Public Support for Protected Areas

David A. Munro

Editor's Introduction:
When *Caring for the Earth* was published in late 1991, it was designed to be the successor to the 1980 *World Conservation Strategy*, which established the powerful link between conservation and development. *Caring for the Earth* describes measures to secure a widespread and deeply held commitment to a new ethic for sustainable living, to translate its principles into practice, and to integrate conservation more effectively with development. In trying to achieve these objectives, governments will find that protected areas play a central role. This chapter, by the director of the team that put together *Caring for the Earth,* provides a useful introduction on building stronger partnerships for supporting protected areas, especially through identifying the nature of the benefits they provide and the persons or groups to whom those benefits are provided. It calls especially for adaptive forms of management that take the various interests into consideration.

Introduction

I use the term "protected areas" in its broadest possible sense in this chapter, because my remarks are relevant to all categories of protected areas, from strict nature reserves to managed resource areas including national parks, wildlife sanctuaries, protected landscapes, and resource reserves. In fact I would like to think that we are talking about the integrated management of public lands for a range of purposes related to human welfare—about a package of activities undertaken to care for the earth. Integrated land management is the process of planning, administering, and managing the use of land in the light of all relevant social, economic, and ecological considerations. It involves the assessment of ecological capability and the evaluation of options in terms of their social and economic consequences. Its goal is the use of land to support a sustainable society. We need to find ways to enlarge and improve public support for comprehensive, integrated land management in protected areas—to expand the constituencies that will see it as being in their own interest to demand that sort of process.

Support for protected areas can take a number of different forms along a continuum from passive to active. Passive support is the absence of opposition. It has relatively little value, but it is better than opposition and it is widespread. It is an important challenge for managers to convert passive support to expressed support, which is far better. It is support from those who are willing to speak out, to stand up and be counted in controversial situations, and to express their views to the political authorities. An important type of support is that which is expressed through vigilance—that is, when nongovernmental organizations and dedicated individuals take it on themselves to keep a watchful eye on the management of protected areas and to raise a hue and cry if it does not seem to conform to legal requirements or established policies. We may also identify supporters who contribute financially by paying user fees or, of perhaps greater significance, by donating funds to create or enlarge parks or contribute to their management.

Why Do People Support Protected Areas?

People support what they believe to be valuable, particularly if they consider it to be threatened or in short supply. People are usually most positive and active in their support if the values that they perceive accrue to themselves. These values may be concrete and easy to quantify, such as the provision of employment or other income; tangible but less easy to put into monetary terms, such as opportunities for recreation; and quite intangible and unquantifiable, such as a wilderness experience. People may also be altruistic and extend support to things that yield them no direct benefits, if they believe that the things they support are of value to other groups or to the broader human community.

The other chapters in this volume describe in some detail the public support for protected areas that may be expected from a number of specific constituencies. In this chapter I will attempt a more general analysis based on the nature of the benefits.

The most clear-cut benefits are those related to employment or income. These may include jobs in protected areas, such as developing and maintaining facilities, ensuring security, providing information, catering to special requirements, and so on. Similar jobs, usually in greater variety, may be created adjacent to the protected area and dependent on it but not subject to its administration. Jobs are not necessarily the only direct source of income that can come from providing services to visitors. If private investments are used as the basis for those services, they may be expected to yield an income to the investors. Most of the benefits related to employment and income are enjoyed by people who live in or near the protected areas, but income from investments may be derived by people from distant points as well.

Other direct sources of income, or of subsistence, may lie in the renewable resources of a protected area—for example, its fish and wildlife and certain plant products. Much opposition to the establishment of a protected area or to its management regime can come from those who suspect that they are going to have to give up traditional uses, such as hunting, fishing, and gathering. However, if arrangements can be made to maintain flows of income or wild goods or to compensate for their reduction or loss, protected areas can expect the support of the people to whom such benefits flow.

Many benefits of protected areas are associated with recreation. Perhaps the most common is a consequence of simply being a visitor, of driving to see points of special natural and scientific interest; of using nature trails, hiking, and climbing; of boating or swimming or enjoying a picnic. These are non-consumptive activities; they require few if any special facilities and usually have a limited impact on the land. Other benefits are associated with recreational activities that require substantial and costly facilities and may have a significant impact on the land. Golfing and alpine skiing in Canada's national parks are examples.

These sorts of uses, besides being a benefit to the participants, also provide employment and income that must be balanced against possible environmental costs. Still other recreational activities, including fishing and hunting, represent a consumptive and possibly disruptive use of a renewable resource, but not necessarily a use that is unsustainable. Recreational benefits are most readily accessible to the people who live nearby, but they attract people from other regions and other countries. Indeed, the existence of protected areas is an important factor in tourism and is, for some countries, a very important contribution to foreign exchange earnings.

Significant benefits may arise from the ecological processes that take place in a protected area. Well known and readily demonstrable are those related to the water cycle. Land that is relatively little disturbed, such as one might expect to be the case in a protected area, naturally slows down the rate of runoff of precipitation, thus conserving water supplies and preventing damaging floods. Undisturbed landscapes, particularly forests, also moderate local climate, reducing extremes of temperature and attracting precipitation. By enabling the management (including protection) of habitat and species, protected areas are major instruments of conservation. For certain species in particular situations, protected areas are not only refuges but also reservoirs: stocks subject to consumptive use outside protected areas may be replenished from within them.

Ecological benefits are obviously most important to the people who are directly affected, such as those who live within the same watershed or enjoy the moderating effects on climate. The conservation benefits are also realized by visitors, particularly those whose purpose in visiting is to see the creatures being conserved. For some others, the conservation benefits are vicarious; they take pleasure simply in knowing that species or ecosystems are being conserved.

As living museums, protected areas can be used effectively in education. The value of learning on the spot is widely recognized, and protected areas provide the most useful and accessible opportunities to demonstrate the functioning of natural and modified ecosystems and to observe diverse forms of life.

Finally, it is fair to say that the country, province, or state that has a well-developed and managed system of protected areas has created an asset that not only attracts visitors and generates economic activity but an asset that is also the basis of a particular prestige, such as is recognized by the World Heritage designation.

How Can Support Be Increased?

We can best approach the question of increasing the support for protected areas by considering the nature of the benefits they provide and the persons or groups to whom those benefits are provided. As a generality, this approach suggests that a goal of the managers of protected area systems should be to increase their actual values by maximizing the benefits and the numbers of people to whom those benefits can be provided—without causing ecological damage now or in the future and at lowest possible cost. It is further apparent that this goal should be sought by better management on the one hand and by raising public awareness of the values and benefits of protected areas on the other.

Management of protected areas is, however, by no means simply a technical matter. Knowing how to ensure that animal populations thrive, and how to build safe roads and lay out good trails, is critical but not sufficient. Managers should identify not only the people who now use their protected areas, but also those who might use them if they were more aware of the benefits they might enjoy or if the management regimes were modified. Protected area managers should know the interests of their publics and assess how well they are being met. This calls for continuing public participation in planning and management.

Actual and potential users of protected areas can be involved in several ways that can usefully be combined. Probably the most important step is to create a protected area council that would draw its members primarily from among those having the most important direct interest in the management of the area. It should meet regularly; it should be kept informed on the uses of the area and the status of its resources; its advice should be sought with respect to any significant changes in management that are being proposed; and it should be encouraged to propose means of improving management. Another method of obtaining public input is by means of periodic enquiries directed at visitors. For this purpose, the traditional technique of distributing questionnaires is useful, and personal interviews of a random sample would be a helpful supplement. Finally, a sample of visitors should be the object of a periodic mail survey to determine changing points of view as well as to reinforce a positive relationship between the protected area and its visitors.

With this sort of regular feedback and always keeping in mind the necessity of managing ecosystems for sustainability, managers of protected areas can base their management practices and planning on sound knowledge. They will be able to undertake the sorts of activities that will be most likely to expand support from their own particular constituencies.

What sort of activities are these likely to be? Planning and management can be clarified if the objectives of each protected area are carefully established and clearly defined. We may assume that it will always be a primary objective to safeguard outstanding examples of a country's or a region's national or cultural heritage. Operational definition of this objective requires a precise statement of what is to be safeguarded by each unit of the system and how it is to be done. Secondary objectives defined in relation to the particular requirements and desires of the constituency served by each unit and aimed at increasing popular support can be aimed at providing the direct and indirect benefits noted earlier. These can be the creation of employment, provision of income, maintenance of traditional patterns of resources use, provision of opportunities for recreation, and so forth.

When possible conflicts between secondary objectives have been resolved, management procedures and projects can be planned and carried out. It may go without saying that management should be to the highest possible environmental standards. But achieving the highest standards need not result in the highest costs. Infrastructure and facilities should be constructed so that they disturb or detract from the environment as little as possible. Without diminishing functional effectiveness, the image should be one of modesty rather than extravagance. Services similarly should be simple rather than excessive; they should minimize waste.

Evaluation is an important component of good management. The information accumulated as a result of contact with visitors is an important input to evaluation and may well serve to alert managers to inadequacies in planning and management. But it needs to be routinely supplemented by carefully designed evaluations aimed at determining the validity of objectives and the effectiveness of measures and procedures aimed at their achievement. And of course evaluations need to be taken seriously; when a change in course is indicated, it should be made.

Conclusion

The creation and good management of protected areas is an essential element in caring for the Earth. Protected areas have an indispensable role in conserving biodiversity. Of equal importance, they can serve as a model of comprehensive, integrated land management, aimed at serving a variety of human purposes within the context of ecological sustainability.

The key to expanding support for protected areas is knowledge of the needs and desires of the publics that are served or affected by the existence of the areas and by the ways that they are managed. Such knowledge, applied within whatever limits may be imposed by the requirement to safeguard natural and cultural heritage, provides the basis for a high standard of planning and management.

Chapter 3

How to Build Local Support for Protected Areas

Walter J. Lusigi

Editor's Introduction:
With vast experience in Africa, Walter Lusigi brings the perspective of a scientist, practitioner, and advocate to the issue of building broader support for protected areas. He begins with a review of the biological need for protected areas and a very important ethical perspective on why broader support is required—a viewpoint especially welcomed from someone speaking with the authority of an employee of the World Bank. He identifies sources of support within local populations, national populations, urban populations, political interest groups, national governments, and international organizations. Calling for mass awareness of the values and function of protected areas and their significance to society, Lusigi stresses the critical role of well-informed protected area managers. In many cases, the responsibility to seek greater partnerships in the wider community rests with these managers.

Introduction

When reflecting on the status and future of protected areas, I have always found it intellectually stimulating and inspiring to look back with admiration at those few individuals who, while conquering territory in a new land, were overwhelmed with what they saw and decided to allocate to humanity what could have been their personal property. Today, numerous dedicated individuals around the world have continued that struggle. From the growth of protected areas to the present network of more than 8500 sites in more than 120 countries, one cannot help feeling some satisfaction and accomplishment. But the battle is far from being won; in fact, protected areas have never been more threatened.

If one compares the expansion of the protected areas network with the expansion of land for agriculture, forestry, irrigation, and human settlements, the picture is not very bright. Likewise the loss of wildlife habitat, reduction in wildlife numbers, and extinction of individual species in the last few decades has been disheartening.

Conservation has always been the idea of a minority. Although the number of sympathizers may have increased, the picture is worrisome in comparison with the population of people who do not care. The environmental crisis that has manifested itself in various disasters like droughts, floods, pollution, desertification, deforestation, and famine has served to focus the world's attention on the environment—this is sometimes confused with attention and concern for protected areas. Whereas millions of people are now worrying about environmental concerns that affect their immediate survival, relatively few are really worried about protected areas. Mitigation measures for various environmental ills like establishment of greenbelts to combat desertification, establishment of grass strips to rehabilitate opencast mining and stabilize sand dunes, protection or restoration of biological diversity and gene pools, sometimes have been mistaken for preservation of protected areas.

Another issue in the debate on increasing support for protected areas is the justification of the functions of protected areas over other concerns and which takes priority over the other in implementation. Four main conventional functions of protected areas are usually put forward: the conservation function, the scientific function, the recreation function, and the education and training function. Which takes precedence in a specific area usually depends on the interest of the viewer. Many have been caught up justifying the existence of protected areas while almost forgetting the fundamental point that the ultimate outcome of the abuse of ecological processes is death.

I would like to suggest here that the most fundamental and basic justification for support of protected areas—which is not necessarily that of the other environmental protection concerns—is that they support life renewal processes that are responsible for the survival of humanity and without which we die. As basic as this might be, it is not always understood even by biologists who work under the different banners. Supporting life renewal processes means providing clean air and water, which requires stable and functioning natural ecological systems that are provided only by protected areas. For a long time I have been working on the management of lands surrounding protected areas because to function as natural systems a minimum critical area is required. This can rarely be provided by the present designated protected areas. There is a minimum area below which a tropical forest cannot function as a tropical forest, and if it cannot function as a tropical forest then it is not a viable protected area and it cannot perform the basic functions mentioned above.

Likewise, protected ecosystems must cater to the year-round survival needs of the animal species found in those systems. In this respect a fauna protected area like those now found mainly in African savanna systems can never be big enough to cater to the year-round grazing needs of animals like elephants and wildebeest. A certain amount of management that includes the surrounding areas is necessary, and this will require the support of surrounding lands and populations. The situation facing wetlands and migratory birds demands even broader cooperation and support, since their habitats and migratory routes traverse large territories.

Why Do We Need Broader Support?

Present systems of protected areas exist because of the vision of our forefathers, who protected and lived with nature. They took from it what they needed for survival and left nature to heal itself. The present nature we are seeking to protect is only a remnant of much wider natural systems that have been disrupted by modern societies whose aim is to find and subdue nature and to find new happiness through industrialization. Almost everyone today, from the loftiest sheik to the most bestial backwoods primitive, has felt some touch of modernity, even if it is only a blown flashbulb discarded by some passing mountaineer or a tin of baked beans from a soldier's mess kit. There is an irrepressible urge among the simpler nations of the earth to possess their own steel mills, computers, and pit-head baths. The technical revolution is seen as the universal panacea, and few people today are prepared to honor the traditional disciplines of society.

In some cases the urge to change is an expression of national pride. But in many developing countries today that urge is an ill-conceived desire, engendered by generations of foreign rule or patronage, to stand all alone in the world, self-sufficient and self-employed. Whereas the politics of sufficiency of food might support the urge for change, in many rural societies there is a longing for a different way of life that is more interesting. Once there has been exposure to pleasures of modernity, life in the rural areas is perceived to be boring. Poverty, natural resource degradation, and increasing populations are good catalysts or justifications for change. Some people have seen change as a more or less irresistible historical process—the closing of one era and the opening of another. Notwithstanding the above, most experts agree that the purely agricultural societies are doomed, and that the impetus of economic development must be toward new industries, new sources of power, and new urban living. This is the changed world that protected areas must live with. To survive they must earn the sympathy and support of all players in the modern development process. This means a change in attitude. In the struggle to meet basic needs in a fast-changing world, the values of natural areas could be easily forgotten. It is the primary responsibility of the protected area managers to ensure that this does not happen.

Sources of Support

Local Populations

In my 1982 address to the Bali Congress on the future directions for the Afrotropical realm, I concluded my remarks with the statement, "Whatever the future of wildlife and protected areas in Africa may be, it is conditional on the support it will receive from the local population." This statement applies to other regions just as much as it applies to Africa. Protected areas are not buildings but rather valuable pieces of landscape found within or among other territories of land use. To continue to survive among these other competing land uses, they must have the acceptance and support of the local populations immediately surrounding them. Although in some situation this support will come from people who understand the values of protected areas, in a majority of the cases the support will only come about if there are no conflicts of use between the protected area and the surrounding areas and if that population is receiving some direct benefit from the protected areas. Some protected areas may not be able to yield direct benefit for the surrounding populations—for example, through tourism or saleable products like game meat and trophies—but may be very important water catchments for the whole region or country. In such cases it would be unfair to expect the local people to bear at their own expense what has come to be regarded as a national or international benefit. The national authorities should work out some mechanism for compensating such communities for the opportunities foregone resulting from such protection activities.

Another aspect of the benefits to the local community should be explained. The compensation of the protected area for lost opportunity should be done in a way to win both material support and emotional support by restoring the long-standing partnership between people and nature. This could be achieved by the returns being invested in community services like education, health, and other social services. Benefits accruing from tourism could also be in the form of employment as tour guides, hotel workers, drivers, and cultural activities like handicrafts and cultural tours. I was impressed by the Australian example, where Uluru National Park has been given back to the Aboriginal people and managed on their behalf by the National Park Service. Such innovative approaches will go a long way in restoring the lost contract with nature.

National Populations

National populations are the next group whose support must be obtained for protected areas. In many developed nations, local tourism is a common thing and the local populations have an opportunity to appreciate the values of protected areas by visiting them. This is not quite the case with developing countries, who see their protected areas as islands for tourism and possibly another way for old colonial powers to maintain a hold on their former territories. Mechanisms need to be worked out whereby the local populations will have an opportunity to appreciate the values of protected areas within their borders.

Urban Populations

Urban populations are an important political and social pressure group in many countries but often have little awareness of protected areas. Because of the nature of the environment in which they live, they may even fear the wild as something hostile. This is an important support group for protected areas; urban people must be educated on the values of protected areas, beginning by making it possible for them to visit these areas. Implementation of such a program will differ from country to country.

Political Interest Groups

Political parties and other interest groups wield a lot of power in decision making at national level. Protected areas should be part of the political agenda in order for the national governments to be adequately sensitized toward conservation. Although many developed countries have achieved this in ringing decrees, in many developing countries the evolving political systems have tended to give only passive consideration to protected areas or environmental issues in general. There is a need to arouse the environmental consciousness at this level by protected area managers taking a keen interest in national political concerns. Here again approaches will differ from country to country but it is quite possible to achieve this consciousness by making conservation components part of welfare programs like agriculture, livestock, and forestry and by stressing the origins of life ingrained in many cultures. This group should also include the business communities that may control governments in some countries.

National Governments

This category of support is sometimes taken for granted, but it must be realized that the lack of support from governments for protected areas has been one of the major limitations in their management. National budgets must be made to reflect the concern for protected areas by balancing them with other priorities like education. Protected areas must be part of the national development program.

International Organizations

International organizations have so far been the backbone of the protected area movement, but often they have been divided in objectives and approaches. Much more can be achieved with the available resources if the international organizations coordinated their efforts better. Although we agree in principle on what needs to be done, in practice institutional allegiances and prejudices have brought about embarrassing conflicts in public, confusing the very constituents we are trying to help. We can do better to mobilize the available financial resources to support protected areas and firmly promote the conservation ethic.

Conclusion

There is no shortcut to obtaining the above support other than through mass awareness of the values and function of protected areas and their significance to society. This mass awareness depends on well-informed managers of protected areas. In my address to the 1986 World Wilderness Congress in Aspen, Colorado, I referred to the need for "a new resources manager," one who is well trained to appreciate the complexity and functioning of present economic and political systems. Although conservation institutions have produced excellent advocates of conservation, we still need lobbyists to ensure that the conservation agenda infiltrates the corridors of power.

Although it is possible to establish greenbelts or plantations that might seem to perform some functions that protected areas provide, it is not possible to recreate protected areas once they are destroyed. Protected areas are natural systems that have a definite structure and function. Maintaining the stability and productivity of these natural ecological systems so that they can continue to provide the life renewal process as they have for thousands of years is what protected area management is all about. Although natural systems have evolved over the years and change is inevitable, it must be allowed to take place within time and space limits that ecosystems can accommodate without destruction. The fundamental truth that must be conveyed to win support for conservation is that the ultimate result of the abuse of natural ecological processes is death. We must repeat this message until all have no alternative but to hear.

Chapter 4

Protection of the Earth's Environment and Corporate Ethics for the Future

Takuya Nagami

Editor's Introduction:
Industry and protected areas are seldom seen as partners. After all, industry is driven by profits while protected areas are driven ultimately by concerns for the welfare of the planet's ecosystems. Yet the undeniable power of industry for both good and bad calls for a thoughtful partnership between protected areas and industry. It is therefore most refreshing that Kaidanren, a powerful Japanese business organization, has taken the lead in developing a "Global Environment Charter," which seeks to provide guidance to Japanese industry on how to make a more productive contribution to environmental issues. Takuya Nagami, the president of Kobe Steel, a major Japanese industrial firm, as well as one of the leaders of Kaidanren, explains how industry can contribute to conservation. Since Japan is a leading industrial power, as well as the world's leading provider of bilateral development assistance, this leadership could be very influential indeed. The challenge for both industry and protected areas is to build on the sound sentiments of the charter and apply them to protected areas. These principles are applicable far beyond Japan.

Introduction

Kaidanren is a Japanese nationwide business organization consisting of 1000 leading Japanese companies and 122 business associations. Keeping close contact with both public and private sectors at home and abroad, Kaidanren endeavors to find practical solutions to economic problems and contribute to the sound development of the economies of Japan and countries around the world.

These days, Kaidanren recognizes environmental problems as major issues in the areas of politics and economics, both at home and abroad. Such problems as limitation of carbon dioxide output, destruction of the ozone layer, acid rain, global warming, the preservation of tropical forests, and the financial aspects of these issues are now regularly addressed by governments and nongovernmental organizations. It is recognized that these problems do not distinguish national boundaries, that they affect rich and poor alike, and that they must be addressed at a global level. Hence the UN Conference on the Environment and Development was held in Rio de Janeiro in June 1992. Mr. Maurice Strong, the Secretary General of the Conference, highlighted Japan's potential role when he said, "Harmonizing the environment and the economy will be the most suitable way for Japan to demonstrate leadership. Japan's technology has been highly acclaimed at home and abroad for its value in antipollution efforts, safeguarding the environment, and saving energy."

It is no exaggeration to say that a company that ignores environmental problems is no longer accepted by society. We at Kaidanren have also recognized that Japan has a responsibility to deal with global environmental problems as a major economic power. These are issues that must be faced by the Japanese business community, which is making strong efforts to disseminate its management system worldwide. Japan's business sector has the will, the potential, and the technology to improve the environment. To meet this challenge, the Global Environment Charter was adopted in April 1991.

Until recently we considered that environmental concerns were antithetical to business in that they seemed to restrict economic growth. Then we realized that, for the long term, such opposition would be against our interest and that business and concern for the environment are not mutually exclusive, but are rather "on the same side." In addition, we realized that it would be better to "come out" first, to be the *proponents* of environmental preservation, rather than be opposed to both governmental and societal pressures to change, passively waiting for regulations to be handed down from government. We too have an interest in preserving the environment, for the present and for future generations. This radical turnaround represents the new spirit of Kaidanren, and the charter was considered with that in mind.

The Contents of the Global Environment Charter

Three main points introduce the charter. First is the necessity of revolutionizing the collective consciousness from so-called "national environmentalism" to "multilateral environmentalism." Japanese companies must also be aware of worldwide environmental problems. We must try to apply our technological expertise in this area.

Second, we must also aim to change both society and the economy, and strive for a leap in consciousness, by cooperating with both the academic and government sectors, and with consumers, in thinking of new ways to apply technology and generally reduce the environmental burden. Third, a positive attitude must be fostered on behalf of the companies, in order to deal seriously with immediate problems that can be solved quickly. In this way, we can establish good relationships with consumers, government administrators, the academic world, and the community as a whole, gaining society's trust.

In seeking to provide guidance to Japanese industry, the Global Environment Charter makes the following points.

- Regarding management policies, companies must bear in mind environmental consequences in all their activities. Notably, the articles of protection of ecosystems reflect an environmental consciousness, an important new point of view.

- The charter also urges companies to set up an internal auditing system, by appointing an executive and creating an organization in charge of environmental activities, which sets up regulations, inspections, and monitoring to ensure compliance.

- Companies should carry out environmental impact assessments of their activities and should utilize technologies that minimize the impact on the environment, and in their procurement activities should promote conservation of existing resources.

- Companies should devote their resources to the development of technology that can be used to preserve natural resources and the environment, should actively engage in transfer of such technology, domestically and overseas, and should consider environmental and antipollution measures when participating in overseas development assistance projects.

- In case of environmental disasters, companies should be open in explaining the full situation to all concerned parties and should take appropriate measures to provide assistance and minimize the impact. They should also actively assist in general disasters, such as those resulting from the Gulf War.

- Companies should participate actively in local community environment preservation programs, educating the public and employees on their activities, providing information on the appropriate disposal and recycling of products and health and safety, and should seek to contribute to environmental policymaking.

- In overseas operations Japanese companies should consider the environmental impact of their activities in the same way as for their domestic operations.

- On global warming, companies should cooperate in research programs and actively work to implement effective measures to conserve energy.

That is a brief summary of the charter's content. We are also interested in the promotion of practical measures to achieve these aims and to develop regional international strategies to address global problems and will be looking into these in the future.

We are very pleased with the reception that the charter has received both domestically and internationally. In Japan it has been favorably received by the relevant sections of society such as the government, labor unions, local government, and consumer groups. Abroad, among other things, it has been recognized as the positive expression of Japanese business toward the ICC Charter on Environmental Issues on the occasion of the Second World Industry Conference for Environmental Management held in Rotterdam in April 1991.

Now, Kaidanren is following up on the charter's impact. For example, we are planning to examine how member companies and organizations are actually dealing with environmental problems. We plan to use our experience to cooperate with developing countries in the conservation of the environment, as it will be helpful for them to learn from Japan's experience in overcoming environmental problems. This could relate to the transfer of Japanese technology for use in environmental and antipollution measures, including the use of ODA (Overseas Development Assistance), and how to put into effect comprehensive measures for educating people. We are also considering other kinds of cooperation. For instance, NGOs (nongovernmental organizations) in the United States, in Central and South America, and in other countries are already engaged in debt-for-nature swaps. To determine how we can help, we have established a group called "The International Environment Cooperative Task Force," and we are now discussing in detail how to transfer environmentally beneficial technology to developing countries, as well as how to implement plans to promote the conservation of nature in developing countries.

Conclusion

In a nutshell, we hold the belief that corporate business must take the leading role by stepping off the rhetorical bandwagon of environmental theory for the sake of good public relations, and instead putting such guidelines to practical use, where the results are visible. Japan has always stressed the idea of harmony. Perhaps we in the East are more attuned to the concept of cohabitation with the environment rather than thinking of it as a resource that man must conquer. We must now show by example how we can live harmoniously with nature. Environmental problems are very complicated and interrelated with many fields. We want to learn the views of experts in many diversified fields who are present at the World Parks Congress, and to have the opportunity to share our views, with the hope of gaining further practical solutions.

Chapter 5

Institutional Options
for Managing Protected Areas

James R. Barborak

Editor's Introduction:
To expand partnerships in support of conservation action, an increasingly diverse and complex array of institutional arrangements is being utilized. In this chapter Jim Barborak, with the Wildlife Conservation Society and the University for Peace in Costa Rica, presents a taxonomy of such arrangements, comparing roles of local, regional, and national government agencies, supranational groups, private landowners, corporations, tribal people, universities, community organizations, nongovernmental conservation groups, and quasi-governmental groups in managing protected areas. Different types of management regimes are suitable for protected areas of different sizes, categories, and importance. Some agencies may own protected areas, while others are responsible for management programs, provide financial support, or have significant input into management decisions. It is apparent that many new institutional arrangements for productive partnerships are being attempted, many of which involve groups that have not traditionally been part of protected area management.

Introduction: A Taxonomy of Institutional Options for Managing Protected Areas

Protected areas vary considerably in their size, diversity, relative local, regional, national, and global significance, and objectives and types of permitted uses and management interventions allowed. Based on these factors, plus the legal and administrative framework for protected area management within a given jurisdiction and land tenure, they also vary radically in regard to the institutional framework for their management.

A taxonomy of protected areas institutional arrangements is not as simple as a dichotomous key. This is because increasingly a partnership involving a number of different public and private institutions is involved in the management of any protected area. However, we can attempt to loosely classify protected areas by answering some of the following questions.

- Who owns the land?
- What is the local, regional, national, and international legal framework for its management?
- Who is responsible for overall management?
- What other players have responsibility for implementing specific management programs or activities?
- Who participates in decision making?
- Who has oversight responsibility to make sure the management agency or landowner manages an area according to agreed principles?
- Who pays the bills?

After such an analysis, most protected areas can be seen as fitting into one of the following general array of institutional options for management.

• *Owned by and at least partially managed by a national government's central bureaucracy.* These areas vary widely in the degree of autonomy of the central bureaucracy, in the extent to which certain park services are entrusted to contractors or concessionaires, or, in the case of research, run by universities and research centers. Individual parks within such a system often also vary in regard to management style, degree of privatization of services, and the role of local communities in management, for historical or legal reasons. Increasingly, nongovernmental groups are helping fund everything from land acquisition through basic operating costs, and are even running endowments, for government-owned protected areas.

• *Owned and managed by parastatal conservation trusts.* This quasi-governmental model is particularly common in the Caribbean; it incorporates the comparative benefits of both public sector oversight and autonomous, private-like administration.

• *Owned and managed by regional authorities such as states and provinces.* These systems are at times truly of global significance but have not yet received the respect they deserve. Particularly in large countries with federal systems of government, state parks are often big, diverse, and in some cases as well or better managed than national parks and reserves. In Argentina, for example, the provincial park system includes 165 areas covering 95,000 sq km; the national park system includes just 23 areas covering 25,800 sq km. In Brazil, the United States, and Canada, provincial and state park systems, while often focusing on providing recreational services, often include sites of international significance and take pressure off fragile national parks; in Australia, the national park system is really a state-run system.

• *Owned and/or managed by local governments such as countries and municipalities.* With a trend toward greater autonomy and power for local governments occurring in many countries, the role of such agencies is bound to grow in protected area management. The role of municipalities is usually limited to managing small areas of local importance for recreation, watershed protection, and similar services, but sometimes, because of strong local interest, municipalities establish protected areas of wider importance, or because government land tenure is at least partially vested at the local level. The municipality of Morales, Guatemala, owns one of the most biologically diverse small pieces of lowland rainforest in northern Central America; when the national government did not act to protect the most important reef for international tourism in the Bay Islands of Honduras, the local municipality established the Sandy Bay Marine Reserve.

• *Owned by tribal peoples.* The growing awareness of the role of tribal peoples in wildlands management is being widely discussed. While some tribally controlled lands are quite degraded, many tribes control important forest, wetland, and coastal areas that if managed as limited-access extractive reserves should definitely be considered part of the global protected areas estate. I would urge our colleagues at WCMC and Cultural Survival to undertake a global review of anthropological reserves.

• *Run by NGOs.* Increasingly, local, regional, and national NGOs are being entrusted with overall management of protected areas, or at least responsibility for some management programs for government parks. Other NGOs are actually buying land and managing it as private reserves. Most private reserves, like Maquipucuna in Ecuador, are small; some are quite large, such as some of The Nature Conservancy's reserves in the United States, and Rio Bravo in Belize, which are tens of thousands of hectares.

• *Owned by private individuals or corporations.* Increasingly, large landholders are allocating all or part of their properties for conservation purposes, including for private game reserves, for sustainable forestry, and for establishing ecotourism lodges. Large corporations in some countries control hundreds of thousands of hectares of forested land, which many times is better managed that similar areas in government hands. As with Category V IUCN areas, such as European protected landscapes, land use restrictions, easements, and other mechanisms may be used to restrict development options over large areas of predominantly private land that maintains important conservation values.

• *Owned/run by universities.* While it is more common for universities to take responsibility for operating research stations or programs at protected areas owned and managed overall by governmental entities, many universities and research institutions own and manage their own individual reserves or reserve systems, which often adjoin much larger government-run reserves. Examples include the University of California reserve system, the Guatemalan biotope system run by the University of San Carlos, the world-renowned research stations including La Selva Biological Station in Costa Rica (run by an international university consortium called the Organization for Tropical Studies), and Barro Colorado Island National Monument in Panama (run by the Smithsonian Institution).

With trends toward bioregional thinking, it is increasingly common to find complexes involving large areas of contiguous land, owned by a range of government, NGO, and private concerns, whose management involves an even greater number of cooperating institutions, and where growing public involvement in the decision-making process makes reaching overall consensus among diverse institutions and individuals quite complicated. One of the greatest challenges of the future for protected area systems will no doubt be how to resolve interinstitutional conflict over management of large bioregions. Such scenarios often involve shared managerial responsibility among a host of agencies with different mandates who must also be open to considerable public input and political scrutiny over the decision making process.

Trends in Institutional Arrangements for Managing Protected Areas

If one reads the regional overview papers on the status of protected areas, and follows trends in how they are managed around the globe, a number of widespread trends in institutional arrangements for protected area management can be identified.

- Greater diversity of institutional arrangements in many nations, at a national level and for individual parks and reserves

- More involvement by individuals living in and around protected areas in their planning and management

- Greater role for nongovernmental conservation organizations and the private sector in general in running protected areas or some aspects of their management, in partnership with national or subnational governments

- More administrative and financial autonomy for individual protected areas, and more regionalization and decentralization in protected area system management

- Larger roles for local and regional governments in protected area management

The institutional hierarchy involved in protected area management is definitely becoming a tangled web of institutional responsibilities in many countries. This responsibility even transcends national limits to involve international organizations through global or regional treaties, conventions, and cooperative programs. In this framework, World Heritage and Ramsar Sites and Biosphere Reserves are the true crown jewels of the global system—areas so important that they have received global recognition of their value. While the institutions involved in supervision of such programs rarely play a day-to-day role in management, their timely intervention can play a major role in strengthening the hand of often-weak national conservation agencies when protected areas are threatened. A good example is the intervention of IUCN and UNESCO in Panama recently, to remind the government of its obligation under the World Heritage Convention, after it had authorized petroleum exploration by Texaco within the La Amistad World Heritage Site. The oil company involved has since withdrawn its plans.

But internationally recognized sites and national parks and equivalent reserves—which can be seen to be the crown jewels of each nation's protected area system—are really just the pinnacle of a global network of tens of thousands of protected areas, from the neighborhood pocket park to the largest transfrontier biosphere reserve. In addition, protected areas can be seen to be an extension of the increasingly common practice of urban and rural land use zoning.

One of the reasons national parks and equivalent reserves have been so successful in developed countries is that they truly exist as the most restrictive form of land use regulation. They form part of a web of different types of protected areas at local, regional, and national levels, each of which provides a wide range of complementary societal services. Many of the problems national parks face in developing countries is that the national parks concept has been sold to developing countries as a free-standing one. Planners have lost sight of the fact that the success of national parks in industrial countries is due at least in part to the existence of a whole range of other techniques for managing natural areas on public and private land to provide goods and services for societies, relieving national parks of that responsibility. However, many developing countries are highly centralized and lack effective regional and local governments, lack political space for the work of conservation NGOs, or fail to provide incentives for production of environmental goods like timber, fuelwood, fish, and other forest, rangeland, and marine products on private lands. In such countries, too heavy a burden is placed on national parks and equivalent reserves for products and services that should be produced in other types of protected areas, through the efforts of private landowners, tribal groups, and management of limited access resources by traditional users.

Major Concerns and Issues for Debate

I would like to point out what I see as some of the biggest potential dangers of the abovementioned trends. Most have to do with a wider concern for excessive pendulum shifts—shifts toward abrogation of government responsibility for protected area management; toward excessive privatization, regionalization, and decentralization; toward abandonment of core areas in the false hope that buffer zone management and attention to local development will reduce threats to strictly protected areas; and toward giving perhaps too much power to local communities over resources of national significance and ownership.

Government Responsibility for Protected Areas
The first concern is the danger that central governments will, by their own decision or under pressure from donors—particularly the IMF and major multilateral and bilateral aid agencies—abandon their role in protected area management under the privatization banner. While restructuring bloated bureaucracies and passing nonvital services to the private sector is now government policy throughout the developing world and in eastern Europe, the thought of total abandonment by governments of their role in managing protected areas of national and international importance gives nightmares to this observer. Central governments have a legitimate role in protected area management even in the most privatized of economies. The setting of policy and standards for protected area management, ultimate approval of management plans, the watchdog role over performance of private contractors and NGOs entrusted to run parks on a day-to-day basis, and the ultimate use of police power and the judicial system in case of blatant threats to protected areas are examples of this role. When governments openly abdicate their role in regulating the actions of the private sector, even in the developed world, bedlam can result.

Excessive Regionalization
When protected area management agencies are part of larger agriculture or forestry ministries, regionalization can lead to chaos when responsibility for

protected areas is taken out of the hands of able protected area technicians in central offices and given to individuals more interested in basic grain production or timber harvest as part of regional offices with wide-ranging responsibilities. In this case, protected areas often get little attention. This has happened in a number of Latin American countries.

Excessive Decentralization

Far too often, park directors have too little say in decision making and often suffer under oppressive bureaucratic constraints. But dramatic pendulum shifts to the other extreme can be even more dangerous, when adequate checks and balances on the actions of local administrators do not exist, and when one person, dominated by local concerns, makes decisions based on political expediency or the urge to make funds quickly off a resource base.

Excessive Local Control over Nationally or Internationally Important Resources

The proposal that all protected areas name a local council to involve local communities in protected area management is very good. The conflict comes when such councils are given ultimate authority or veto power over management decisions for protected areas of national or global significance and on national lands. For example, should a local council have the right to permit forest cutting to improve the local tax base if this will affect water flows to downstream hydropower or potable water projects benefiting an entire region? A good example of this situation occurs with grazing permits on Bureau of Land Management Lands in the western United States.

Losing Sight of One's Mission

One of my biggest concerns is that with all the talk about buffer zones, extractive reserves, sustainable development, and the need to improve the lot of local peoples in regions surrounding protected areas, the institutions involved in management might lose sight of their core mission, which in the case of strictly protected areas is to protect resources for the long term and produce a sustainable flow of wildland services for a nation. Protected area agencies are jumping on the sustainable development bandwagon and diverting scarce human and financial resources to tackle problems they can never hope to resolve. Many such problems are really the responsibility of other agencies, and this diversion of interest is occurring at a moment when most protected area agencies are suffering cutbacks or frozen budgets and staff levels in an era of growing responsibilities. What good is a buffer zone if the core is destroyed?

The Pressure of Aid Agencies to Downsize Government and Its Impact on Protected Areas

The IMF, IDB, World Bank, and other sources of finance are rightfully insisting on fiscal and bureaucratic reforms as conditions for helping countries make the transition to open economies fully inserted in the world system. Their efforts to insist on reducing fat in government, holding back government spending, and reducing payrolls are good in general. However, their unbridled enthusiasm about the ability of the private sector to work more efficiently than government, and their obvious trend toward funding private initiatives over public ones, is of concern. Unfortunately, their policies have hit weak conservation agencies particularly hard in a decade in which the number of protected areas has risen dramatically.

Although privatizing some park services and giving more responsibility to NGOs and local communities for protected area management can take some of the pressure off central government, the truth is that economic restructuring agreements have devastated many park systems. Particularly tragic has been the loss of many gifted civil servants, from rangers to park system directors, who have jumped to the private sector due to poor working conditions, often lured by incentive packages now being given to voluntarily leave government.

Donors have to learn that progress in parks is not to be gained by robbing Peter to pay Paul, but rather by strengthening both public and private sector institutions and better defining the appropriate niches for a whole range of institutions in protected area management. This should be done without losing sight of the legitimate role of government in areas such as defining management norms and regulations, setting overall policy, playing a watchdog role over NGOs entrusted with management of public lands, and the legal and police functions. I remain convinced that government protected area agencies, given sufficient financial and administrative autonomy and freedom from gross political intervention in day-to-day management, can and do function quite efficiently.

Conclusion

To sum up, increasingly diverse and complex arrays of institutional arrangements are being used to manage protected areas in many nations, with greater involvement of local peoples, NGOs, private landowners, corporations, subnational governments, and universities. Many such protected areas managed by subnational governments, tribal peoples, or private entities protect resources of national or global importance and as such should be considered part of the global protected area network. There is no doubt that stronger roles for all the abovementioned actors, and development of protected area systems where national parks are truly crown jewels and the pinnacles of protection in complex park and reserve systems, will lead to strengthened national and global protected area networks, capable of providing a much wider and more sustainable flow of environmental goods and services to society.

Chapter 6 ─────────────────────────────

Bringing in the Law: Legal Strategies for Integrating Habitat Conservation into Land Use Planning and Management

Parvez Hassan

Editor's Introduction:
Many of those involved in protected area management may feel that the less they have to do with lawyers, the happier they will be. But Parvez Hassan, chairman of IUCN's Commission on Environmental Law, shows that a partnership with the legal profession is essential to enabling protected areas to meet their conservation objectives. He describes a number of legal techniques for conservation on both public lands and private lands, demonstrating that many tools are available to protected area managers, private landowners, NGOs, and many others interested in achieving conservation objectives. While the legal instruments required to integrate conservation with development are still in their infancy, the further development of this essential foundation for conservation will be built on a strong partnership between lawyers, managers, and practitioners concerned about specific protected areas, protected area systems, and the global legal framework that supports them.

Introduction

It has now become an accepted fact among conservationists that, necessary as they may be, protected areas will never be sufficient in themselves to conserve the full range of biological diversity. Therefore, other means are also needed to reach this objective. A number of countries, mostly in the developed world, have during the past several years developed for that purpose a variety of innovative legal instruments of a regulatory or voluntary nature that constitute a reservoir of ideas and techniques, or even useful models, for other countries.

Events particularly in the past two decades in the developing world have shown the importance of a constitutional commitment to environmental protection for the conservation of both public and private lands. The developed world has utilized a plethora of specific subconstitutional legal instruments to achieve conservation objectives. The developing countries have lagged behind in such responses. But the apathy of the parliaments in several developing countries has been remedied by the vision and activist role of the judiciary in these countries. Even very general pronouncements in the constitution have enabled judges to grant relief to the public interest groups beyond the wildest expectations and dreams of the draftsmen of these provisions. Thus, for example, in India, a general provision on the right to life was interpreted by its Supreme Court to mean a right to dignity and a clean environment, and was the basis to strike down stone-quarrying and mining rights in national parks and other protected areas. Similarly, a written petition filed in Pakistan to protect the Margallah Hills National Park has led to recent action by the prime minister to prohibit the protested use of the national park. A commitment to include provisions on environmental protection in the constitutions of particularly the developing countries will both influence and facilitate judicial intervention to respond to the efforts of public interest groups and guide the preparation of national development plans.

Legal Techniques for Conservation on Public Lands

Vast areas of land are still in public ownership in many parts of the world, of which protected areas constitute only a small fraction. Conservation of natural areas on public lands may be achieved by transfer of land to a conservation agency and the establishment of conventional protected areas on such land. However, it may not always be possible to do so without specific authorization by an act of parliament. In addition, rules relating to the transfer of public property from one agency to another often require that land be sold to the other agency at its market price. Finally, where land is held by an agency for a specific public purpose, it should remain in the hands of its managing agency. In many countries the conservation agency is the forest department, and there may be a natural reluctance on the part of that department to withdraw some of its lands from exploitation.

Ways must be found to ensure that the conservation of important natural habitats in public ownership be effected by the land-managing agencies themselves. One possibility is to enact legislation providing for the establishment of wilderness areas, large tracts of public lands that have remained in a pristine or quasi-pristine condition and where the construction of roads or tracks and the use of motor vehicles are prohibited. The concept was developed in the United States by the Wilderness Act of 1964 and has been since that time taken up by several other countries, such as Finland, New Zealand, and Sri Lanka.

Another means, which relates more specifically to public forests, consists in developing legislation requiring forest departments to maintain the multiple functions of natural forests and to seek to achieve a reasonable balance between timber production and the conservation of biological diversity. A certain number of modern forest acts contain provisions along those lines. The drawing up of multipurpose forest management plans and the establishment of environmental impact assessment requirements for forest policies, programs, and major projects may facilitate these objectives.

It is also possible to provide that a conservation agency may, by regulation, impose restrictions on the use of land held by another agency, or to empower it to conclude management agreements with other agencies for conservation purposes.

Finally, landholding agencies may be encouraged by legislation or by agreement with the conservation department to set up their own networks of protected areas on their lands, to develop a capacity to manage them themselves for conservation, or to accept that they be managed, under contract or otherwise, by the conservation agency or by some other organization.

The acquisition of land for conservation by government agencies is also an essential instrument for the preservation of natural areas. Legal means to facilitate acquisition may be necessary, for example. One of these is to provide for a right of preemption (or right of first refusal) regarding land coming on the market. Another is the right to purchase land compulsorily for conservation purposes. There are still some countries where this right does not exist.

Legal Techniques for Conservation on Private Lands

Conservation on private lands may be achieved by regulatory or voluntary measures or by a combination of these measures.

Regulatory Measures
Regulatory measures are designed to impose restrictions to the rights of private landowners in the interest of conservation. Such measures usually apply to particular sites or landscape features. They often take the form of conservation orders prohibiting or limiting certain activities in certain areas or protecting certain landscape elements such as individual trees or groups of trees.

A unique system is the Sites of Special Scientific Interest in Great Britain. The Nature Conservancy Councils have a statutory duty to designate sites that in their opinion are of such interest. Designations have a certain number of legal consequences. In particular, landowners are notified of the activities, including farming activities, which they may not undertake without first informing the appropriate council of their intention. The council then must negotiate a management agreement. The interest of the scheme is twofold: It is the scientific interest of a site that is the source of its legal regime; and it is one of the few examples where regulatory and voluntary measures are used in combination.

Another kind of regulatory measure consists in requiring a permit for the destruction or alteration of certain habitat types. This system is applied to wetlands or to certain categories of wetlands in a number of jurisdictions—in particular, in the United States, where it exists both at federal level and in many of the states.

Another very successful scheme is the one instituted in the Australian state of South Australia, where a permit is required for the clearing of any area of indigenous vegetation.

Regulatory measures applied to private lands for the purpose of conservation are generally considered a legitimate use of the police power of the state in the public interest. Compensation is, however, generally due when regulating measures are site specific. Whether they are also due when restrictions apply to all lands in a certain category, as would be the case with respect to habitat types, is a matter of controversy. Another major drawback of purely regulatory methods is that they cannot normally provide for the management of protected habitats unless special agreements are concluded with the owners for that purpose. For these reasons nonregulatory methods of conservation are becoming increasingly popular.

Voluntary Measures

Voluntary conservation may be achieved through the encouragement of unilateral commitments by landowners, the conclusion of contracts, and financial incentives or disincentives of various kinds.

Landowners, including conservation NGOs, are usually entitled to establish private reserves on their land. These reserves usually have no specific legal status. In a few countries, such as France, voluntary reserves, after approval by the Ministry of the Environment, benefit from the same degree of protection as statutory nature reserves. Private reserves also need to be protected, as far as possible, from compulsory sales to public authorities for purposes other than conservation. In the United Kingdom, the National Trust, a major owner of land held in trust for conservation, has the right to declare its real property inalienable. This may be reversed only by a decision of parliament. Similar systems exist in certain states of the United States where landowners have a right to "dedicate" land to conservation.

Financial incentives are another means to encourage the creation of reserves on private lands. In Finland the law does not allow the establishment of reserves on such lands without the consent of the owners. The latter are, however, entitled to receive payments from the appropriate authority when they have given their consent. In the Netherlands, NGOs benefit from substantial grants for the purchase and management of land for conservation.

The conclusion of conservation contracts between conservation agencies (or sometimes conservation NGOs) and private landowners has now become an established practice in an increasing number of countries. Such contracts usually do not provide for positive management obligations. Management agreements whereby landowners undertake to manage their land in a specified way in exchange for regular payments are a solution to this problem.

It is also possible to use various forms of financial incentives or disincentives to encourage the preservation of the natural environment by private landowners. These include special subsidies or tax exemptions to farmers, as well as the removal of certain tax benefits, subsidies, or other incentives to the destruction of natural areas. While the payment of conservation subsidies is usually linked to the conclusion of a management agreement, tax exemptions can apply to a variety of situations, including the transfer of land to government agencies or conservation NGOs for conservation purposes or the maintenance of an area of land in its natural condition.

A recent example of disincentives is provided by the United States legislation that makes farmers ineligible for a large number of federal benefits, including price support payments, federal crop insurance, and federally insured or guaranteed loans, if farmers produce agricultural commodities on highly erodible land or on wetlands.

All the measures described so far are sectoral and only apply to particular situations. It is, however, increasingly recognized that conservation measures should be integrated into land use planning and that the countryside should be managed in a comprehensive way, all ecological factors being taken into consideration. This requires new approaches to land use legislation.

Legal Techniques for the Integration of Conservation and Land Use Planning

One possible method is to provide, as does Swiss law, that natural areas must be included in the zones of local land use plans where restrictions on development are the greatest. As such restrictions usually do not apply to agriculture and forestry, this may not be sufficient to conserve all areas worthy of protection; additional regulations may be necessary to prohibit activities that may be detrimental to the integrity of the areas concerned.

It is also possible to designate areas where more stringent land use restrictions than in the rest of the country are to be applied. In France, for instance, special rules have been enacted to preserve natural habitats in coastal and mountain regions through the planning system. Alternatively, the law may decide that all activities that may have adverse effects on the environment shall be controlled.

However well devised the legal system of conservation of areas may be, it will remain inefficient if its usefulness to individuals and communities living in or in the vicinity of these areas is not clearly perceived. This is particularly the case in developing countries, where conservation measures have in the past often been perceived as running counter to the immediate well-being of people. The challenge is to provide local communities with sufficient incentives and disincentives to encourage them to respect the restrictions established by law; this requires different approaches in different situations. A particular challenge is to consider how local communities can participate in the establishment of these restrictions and in benefits derived from them.

Conclusion

Unless the needs of the common people are met, legislation, however well-intentioned, will remain unimplemented and conservation will remain an elusive goal. The important conclusion is that the integration of conservation and development and the legal instruments required to achieve this end are still in their infancy. Yet if enough is to be saved of the biological diversity of our planet, it is essential that adequate and effective legal means be developed, tested, and, where necessary, adapted to the specific requirements and situations of all countries in the world. There is a need, therefore, to share information on innovative legal conservation techniques and of their effectiveness in ensuring or promoting conservation, and to promote the establishment of a continuous dialogue on these issues between lawyers, managers, and practitioners from all parts of the world.

Chapter 7 ————————————————————————————————

Social Science and Protected Area Management: The Principles of Partnership

Gary E. Machlis

Editor's Introduction:
While most protected area managers have backgrounds in biological sciences or forestry, it is becoming increasingly clear that the management of protected areas in the 21st century is necessarily the management of people. And managing people is a difficult task that will be facilitated through the use of the social sciences at the protected area, regional, national, and global levels. The social sciences of anthropology, economics, geography, psychology, political science, and sociology are all highly relevant to protected areas, and a partnership between protected areas and social sciences can significantly enhance the role of protected areas in sustaining society. Social scientists can help provide feedback, prediction, assessment, and mitigation techniques for protected area managers. Gary Machlis, from the U.S. National Park Service and the University of Idaho, provides a number of recommendations for an invigorated partnership between social sciences and protected areas, calling for effective monitoring programs, an international network of cooperative protected area studies units, integrating social science research programs into natural science research programs, and using existing bureaucracy creatively to encourage the production and application of usable knowledge.

Introduction

The management of protected areas is necessarily the management of people, for kin, community, class, and culture are fundamental units in the use, conservation, and preservation of natural resources. The past decade has seen a growing realization within the conservation movement that biophysical and social systems are inextricably intertwined. Hence, the social sciences have emerged as a potential partner to conservation in general and to protected area management in particular. As the theme of the IVth World Congress is enhancing the role of protected areas in sustaining society, the social sciences and protected area management seem poised for important cooperation. This chapter describes this partnership and makes recommendations for improvement.

Several questions guide the analysis: What do protected area managers need that social science might provide? What exactly have the social sciences contributed that is "usable knowledge" for protected area managers? What contributions can be expected in the future? What is required to enable the social sciences to become an integral part of the protected area movement? The answers attempted here are personal and subjective; other social scientists would likely provide different views and opinions. While the scope is international, the limits of language result in a general reliance on my experience with the English-language scientific literature. An overview and synthesis is intended, rather than a review of research results. The recommendations are hopefully significant and amenable to action.

What is meant by "social science"? While definitions vary (and often confuse rather than clarify), the key characteristic of social science is the application of the scientific method to understanding social behavior. Those academic disciplines that include significant amounts of social science are anthropology, economics, geography, psychology, political science, and sociology. The distinctions between "real science" and social science, or between the "hard" and "soft" sciences are largely intellectual marking of territory and of little importance: There is really only the scientific method, poorly or well applied. Nor is one or the other social science necessarily preeminent; all have potential to contribute to conservation.

Usable Knowledge and the Principles of Partnership

Protected area managers are faced with an often-bewildering and complex set of decisions, most of which must be made relatively quickly, simultaneously, without complete information or understanding, and with feedback effects that then must also be dealt with by additional decision making. A majority of these decisions have a socioeconomic or sociopolitical component: Actions to be taken will likely have important impacts on the wider social system. Hence there is an almost continual opportunity for social science to assist in making such decisions, if it can provide "usable knowledge." The criteria for usable knowledge related to protected area decision making are specific.

- The information must be provided at the proper point in the decision-making process. Timeliness is critical.

- The information must directly address the manager's "need to know," and at a level of detail appropriate to the decision.

- The manager must understand the limitations of the data, the degree to which it can be applied, the certainty (or lack thereof) of successful application, and the authoritativeness of the authors.

Hence, a research project completed too late, dealing with issues of only tangential relevance to a manager's decision-making needs, presented without limits or explanation and by scientists of unknown credibility, will not likely result in usable knowledge. Note that such research could be excellent, even brilliant science; it would still remain outside the boundaries of usable knowledge. A first principle for organizing an effective partnership can thus be stated: The social sciences must provide usable knowledge to protected area managers.

While the decision-making activities of protected area managers are often undertaken within a complex sociopolitical context, the use of scientific information in such decision making is, in reality, quite limited. Information from the biophysical sciences are more likely to be employed than that from the social sciences; a water quality assessment or game population estimate is more likely to enter into a resource management decision than an employee survey into an administrative one. Protected area managers often use common sense, folk knowledge, field experience, and ideological views to make decisions, and usable knowledge from the social sciences is frequently ignored or avoided.

In many cases, managers may not be aware of or understand the potential advantage of using social science information. Often, protected area managers are uncomfortable integrating scientific information into their decision making. Scientific advice often limits the range of decision alternatives available to the manager, by identifying unacceptable consequences, prioritizing choices along scientific rather than political criteria, and creating the need for managers to defend their rationale for not following such delivered advice. For all these reasons, what currently occurs is ad hoc and fragmented use of social science information. Its potential is not being fully exploited. A second organizing principle for a full, effective partnership can thus be stated: Protected area managers must integrate the usable knowledge of social science into decision making. How such integration might realistically occur, and to what degree protected area managers might profit from using social science, is discussed below.

The Critical Importance of Scale

Protected area management takes place at significantly different scales, and the issue of scale is central to the partnership of science (social and biological) and conservation. Table 7-1 illustrates the major scales of protected area management. For each, key organizational units are considered in decision making. At the *protected area* level, key units of organization include visitor groups, resident populations, park staff, and within-park enterprises. At the *bioregion* level, the protected area is seen as embedded in a wider ecological and social system, with boundaries conceptually defined rather than gazetted.

Regional units of concern include local communities, states and provinces, regional offices of park and other natural resource agencies, regional markets, and service economies. At the *national* level, key units are the national legislatures, central administrations, large nongovernmental organizations (NGOs), the media, and other national agencies managing resources. At the *realm* level, international organizations and other nations' park agencies are

Table 7-1. Scales of Protected Area Management and Key Organizational Units

Scale of protected area system	Key organizational units
Protected area	Visitor groups
	Resident populations
	Park staff
	Concessions
Bioregion	Local communities
	States and provinces
	Regional offices
	Regional service economies
National protected area system	National legislatures
	Central park administrations
	National NGOs
	National travel industries
	Bilateral NGOs
Realm	International NGOs
	International treaty organizations
Global system	National NGOs
	International travel industry
	United Nations

central. At the emerging *global* level, international NGOs, treaty organizations, and world markets become significant organizational units.

At each scale, the decision-making process of protected area managers will vary, since different organizational units and political contexts interact. That is, the management of protected areas is scale-dependent. In addition, each level of management is significantly influenced by the adjacent levels, and all are in actuality parts of a nested system of protected area management. Information needs of protected area managers will differ at each scale, though contributing to an overall set of needs. Hence, what will be considered usable knowledge at one scale may be irrelevant or of little use at another.

Table 7-2 illustrates this idea of scale dependency. At each scale, several primary ecosystem and institutional issues are suggested. Each is linked to management issues at other scales. For example, habitat change and population loss at the protected area level can contribute to habitat fragmentation and species loss at the regional level; policy formation is a major institutional issue at the national level and is a significant component of strategic planning and international cooperation at the global level. So significant is the scale

Table 7-2. Key Issues by Scale of Management

Scale of protected area system	Key ecosystem issues	Key institutional issues
Protected area	Population loss Habitat conversion Exotic introductions Ecosystem effects	Visitor services Resource management Sustainability Local populations
Bioregion	Habitat fragmentation Species loss Ecosystem stress	Training Monitoring Coordination Policy implementation
National protected area system	Reduced biodiversity Species loss	Policy formation Funding Acquisition Development strategies
Realm	Reduced biodiversity Species loss	International cooperation
Global system	Reduced biodiversity Climate change	International cooperation and assistance Strategic planning

dependency of protected area management that concern for differences between small and large areas, urban and rural, developed and developing countries, north and south can be reasonably suspended. Since these scales largely determine the social science information needs of protected area managers, each level will be discussed in turn.

Protected Area

Three organizational units predominate at the protected area level: visitor groups, resident populations, and employees. Managers at the protected area level need to document the social ecology of visitors—that is, the relationship of visitors to the park environment. Their distribution, abundance, demographic composition, behaviors, and resource demands are all important variables in determining ecosystem impacts and viable resource management strategies. Visitor wants, needs, opinions, and expenditure patterns are valuable in policy and marketing decisions. To be useful, such information must be contemporary, area-specific, and, where visitation varies by season, season-specific. In addition, managers need ways to predict changes in visitor use, effectively manage visitor services, design efficient facilities, and readily communicate protected area values to visitors.

Resident populations present managers at this level with a different set of information needs. The numbers, distribution, and demographic composition of resident populations are of course important. In addition, managers need to understand sustenance and cultural requirements of such peoples, and their impact on protected area resources. Information must be area-specific, accurate, and sensitive to cultural differences. Managers need strategies for coordinating decision making with resident political structures, and for setting sustainable levels of resident economic activity while protecting park values.

Employees are also a crucial organizational unit, and at the protected area level, several information needs emerge. Employee job satisfaction, morale, and concerns should be monitored as a feedback mechanism for improved administration. The information must be area-specific, accurate, and timely. Managers need effective supervision, training, and staff development techniques.

Bioregion

While protected areas are largely defined by their legal and/or political boundaries, protected area regions include the protected area and adjacent, related ecosystems and human communities. Biosphere reserves are an exception, being essentially institutionalized protected area regions. Several organizational units are crucial to the management of such regions, and present managers at the regional level with a unique set of concerns. Local communities, particularly those at or near gateways to protected areas, produce several information needs. These include an understanding of population trends and economic activity levels; a grasp of critical cultural values, political structures, and leadership processes; and the dependency of such communities on park and regional resources. Assessment of sustainable development levels, prediction of social and economic impacts of policy decisions, and strategies for effective public involvement are all valuable management tools.

Another example are regional governmental institutions, particularly regional and provincial governments. Here, managers need an understanding of regional political processes (both ideal and real), power-sharing arrangements (both formal and informal), and agency decision making. Since protected areas are increasingly used as tools for economic development, knowledge of regional economic trends (including labor and capital flows) is both valuable and necessary. Strategies for evaluating the social and economic impacts of regional development projects and for interagency coordination of governmental activities are needed.

Nation

Managers at the national level are faced with yet another and different set of organizational units. National legislatures, central agency administrations, national NGOs, media and industrial sectors (such as the tourism industry) are examples. Information needs vary dramatically from previous levels. For example, while area managers need specific, seasonal descriptive information about park visitors, national managers do not. National managers need accurate statistics on total visitation levels, including trends, future projections, and, to a lesser extent, regional distributions. Data on the economic impact of protected areas are politically valuable, as are techniques for predicting future trends in visitor use and principles for design of standardized facilities and services.

Administration is a central concern at the national level, and information required for effective administration includes staffing requirements (both current and projected), inventory of human and financial resources, and evaluation of subordinate managers. Techniques for allocating scarce resources, monitoring the status of individual protected areas and regions, and training and supervision of employees are all required at this level. National policy initiatives, head-of-state decisions, and media influence are crucial elements in decision making, and the ability to conduct policy analysis, respond to executive information requests, and monitor public opinion is both valued and necessary.

Realm

Managing protected areas at the realm level is an example of emerging scale, and fewer kinds of organizational units have evolved than at the other levels so far described. International NGOs, bilateral cooperative ventures (through treaty, contract, or agreement), and nascent realm organizations (such as that within IUCN's CNPPA) are examples. Management largely involves strategic planning, monitoring, training, the administration of international aid programs and technical assistance. Information needs include assessment of research and development applications, monitoring of critical problems (both general and endemic to the realm) either at the national or area level, and assessment of technical assistance needs. Strategies for improving the efficacy of technical assistance programs, enhancing the adoption and diffusion of innovations, increasing communication between national-level managers, and networking among NGOs are significant needs.

Global System

Like realm management, global system management is an emerging scale in conservation, and particularly in protected area management. Organizational units include the United Nations (and its subsidiary institutions), IUCN (and its subsidiary commissions), the globally operating NGOs (such as WWF), and national NGOs with international agendas. Also included are the developed nations' donor agencies, and world trade associations related to travel, tourism, and natural resource production. Management tasks revolve around strategic planning, allocation of resources, and technical assistance. Hence information needs of these managers tend to be monitoring of global trends (often using national-level data) and policy analysis. The ability to provide documentation and support for global initiatives, as well as to assess the viability of conservation strategies within different social, political, and economic systems are paramount needs of managers at this level.

The scale dependency of protected area management creates a wide range of information needs that can be addressed by the social sciences. However, it is not realistic to expect all of the social sciences to contribute equally to usable knowledge at each management scale. The social sciences diverge according to their key units of analysis, central concerns, and experience in protected area management issues.

The Contributions of Social Science
to Protected Area Management

Since the early 1970s, the social sciences have made considerable progress in their understanding of issues related to conservation generally and to protected area management specifically. An example is in economics, where concepts such as maximum sustainable yield and marginal opportunity cost have been employed to better grasp the causes and consequences of natural resource production. Literally hundreds of articles, essays, research reports, and books are published worldwide each month. The contributions of usable knowledge are, however, more modest. Numerous social scientists are working on specific projects that have had or will produce useful results; their work is admirable and indicative of the social sciences' potential. If, however, we move from individuals to more widespread contributions—that is, search for a pattern of sustained usable knowledge—the results are meager and frustrating. Some examples, organized by the scale of protected area management, are described below.

At the protected area level, most usable knowledge has been the result of applying social science research techniques rather than theoretical understanding or prediction. Visitor surveys have become common, though they are irregularly taken, often poorly designed and administered, and seldom archived for future use as baseline data. Protected area managers have used survey results to "better understand" their visitors, establish the economic impact of tourism, and evaluate visitor services. Their use in decision making has been largely limited to influencing minor policy changes and facility design. Several techniques for limiting or planning visitor use have been adopted by protected area management agencies, derived from an amalgam of social science theory (primarily psychology) and field studies.

At the regional level, several of the social sciences (particularly anthropology and sociology) have provided protected area managers with usable knowledge regarding local populations and communities. The results, usually detailed cultural descriptions, have increasingly been integrated into decision making by donor agencies and technical assistance programs, and, to a lesser extent, protected area planning. Economic analyses have in recent years begun to provide input into the strategic planning of sustainable development; since "sustainability" takes years to assess, the value of such inputs remains to be seen.

At the national level, the contributions of usable knowledge are especially sparse. Some basic data collection is continuous at this level but of relatively chaotic quality; it is most often used by the media and in budget justifications. Economic measurement of protected area economic activity has been visible, yet its integration into decision making is primarily through the political system, as leadership groups vie for dominance over resources on marginal, public, or communal lands.

At the realm and global levels, social science has provided a minor but growing contribution. Monitoring of global trends (primarily biological, but including social indicators such as per capita income, population growth and so forth) has become popular, though its actual use in decision making is unclear. Geographic Information Systems (GIS) technology is now being applied at realm and global scales and has been useful in the allocation of resources (particularly during emergencies such as drought). In some respects, the work of anthropologists, geographers, economists, sociologists, and others has documented the need to link protected area management and the sustainability of local peoples, leading to a new paradigm of protected area management and directly contributing to the theme of the IVth Congress on National Parks and Protected Areas.

The Potential of Partnership

While my assessment of the current partnership of protected area management and social science has been harsh, the potential contributions of usable knowledge give cause for enthusiasm. The social sciences can provide considerable usable knowledge, if properly focused and organized. Protected area managers can integrate such information into their decision making, if properly prepared. And such a partnership can enhance the role of protected areas in sustaining society.

From a systems perspective, the most valuable contribution of the social sciences may be classified as feedback and prediction. The major uses of these by protected area managers in decision making are for assessment, and mitigation. Feedback, prediction, assessment, and mitigation form the core of partnership across the scales of protected area management.

At all scales, the social sciences can and should focus on developing feedback mechanisms for managers. Visitor surveys, monitoring of resident population resource needs, and reporting of socioeconomic trends are examples of important feedback activity. The requirements of usable knowledge demand that such feedback be timely, deal with trends important to managers, and have clear and demonstrable scientific integrity. Social scientists must therefore focus on adapting all aspects of their research techniques to the practical needs of managers, from study design to the final reporting of results.

The role and importance of prediction in science cannot be overstated. Prediction is the essence of the scientific method, and hence good science must attempt and provide prediction. Social scientists working on protected area issues have for too long avoided prediction for the safer realm of description—describing in social science terms what managers often see for themselves. The storehouse of theory and prediction available from the social sciences needs to be opened up to protected area managers. Social scientists need to apply their theories and make concrete predictions—about sustainable activities, biodiversity loss, visitor satisfaction, cost and benefit, and a host of other managerial concerns. These predictions should be based on tested theory rather than favored ideologies, and the level of certainty assigned to each prediction must clearly be described. Some predictions will undoubtedly turn out in error; such results can be used to improve future predictions. When a protected area manager asks, "What will happen if I do X?," the social sciences must attempt an answer.

If the social sciences provide usable knowledge in the form of feedback and prediction, then protected area managers have a real opportunity to integrate such knowledge into their decision making. One important arena is assessment. However informal, most protected area managers attempt an assessment of conditions prior to making decisions, from the siting of new tourist facilities to the regulation of subsistence use. Managers need to build into their assessments a role for social science information. The more formal their assessment process (which will vary by scale, importance of decision, and other factors), the more formal a role for social science is required. For example, protected area planning should include a significant level of social science information on visitor, resident, and nearby population resource needs, and the planning process should be designed to make this possible.

In addition to using social science in assessment, protected area managers will benefit by employing such expertise in the mitigation of impacts. Protected area management decisions have consequences intended and unintended; a new visitor road opens up an area for poaching, or a new regulation leads to conflict between locals and tourists. Armed with the predictions of its partner social science, the protected area manager at all scales can better mitigate effects. Social science can provide, if managers are willing, useful strategies for dealing with the consequences of decisions. Examples include the use of economic incentives, communication techniques, and conflict resolution.

These functions—feedback, prediction, assessment, and mitigation—form the core of a successful partnership between social science and protected area management. What institutional change is required to achieve such cooperation?

Conclusion: Recommendations for an Invigorated Partnership

Institutional arrangements have a great influence on how social science and protected area management can and will cooperate. While there are significant differences in the level of partnership throughout the world, and at the different scales of protected area management, some general actions can be proposed.

• *At each scale of protected area management, monitoring programs should be established.* Some programs exist: Many protected areas keep track of the number of visitors, and PADU (the World Conservation Monitoring Center's Protected Areas Data Unit) represents an important effort at the global level. Yet systematic monitoring of socioeconomic trends is currently not available. Social scientists should develop these programs, and managers should be involved in determining what data are collected. Feedback to managers should be continuous and in easy-to-use form. Data collected at one level should, as much as is possible, be aggregated at the next. For example, national-level data can be combined to form indicators of realmwide conditions. A major global assessment of key socioeconomic trends should be produced prior to each World Congress, beginning in 2002.

• *An international network of Cooperative Protected Area Studies Units (CPASU) should be established.* These research facilities should be located (whenever possible) at universities and funded by protected area agencies, and should employ a mix of university and agency scientists. Such units are a viable and efficient way of producing usable knowledge in both the social and biological sciences. First institutionalized in the Pacific Northwest Region of the U.S. National Park Service, CPASUs can and should be adapted to the particular needs of each region, country, and realm. To staff such units, a generation of young, home-country social scientists must be nurtured and encouraged to apply their skills to protected area management. A network of such research stations can play a major role in the monitoring described above.

• *Social science research programs must be integrated into natural science research programs.* One of the barriers to the full use of social science by protected area managers has been that social science has most often been treated separately from the biological sciences in funding, staffing, and organizational structures. Since the problems faced by protected area managers are interdisciplinary, this artificial separation has led to a host of problems: lack of cooperation between biological and social scientists; inadequate and undependable funding for social science; lower standards of scientific rigor; and, most important, reduced usable knowledge for managers. While integration of the sciences will not solve all these problems, it is a necessary precursor to significant improvement.

• *Existing bureaucracy must be creatively used to encourage the production and application of usable knowledge.* In many cases, existing bureaucratic structures, regulations, and policies have the potential to encourage and increase the amount of usable knowledge produced and used. For example, much of current social science is conducted under contracts or formal agreements between researcher and protected area agency or organization. Such contracts can, if carefully prepared, increase usable knowledge by requiring manager involvement in study design, stipulating the need and format for usable results, and including as necessary products training workshops for managers on how to use the research in decision making. Likewise, current supervisory systems can be revised to create incentives for managers to integrate social science into their decision making, by requiring formal assessments, evaluating managers on their use of social science in relevant decision making, and by a significant increase in relevant training.

Other recommendations are certainly appropriate, and these can be improved on. Note that I have not made the generic and expected recommendation that funding for social science be dramatically increased; a long-term strategy for partnership suggests that increased efficiency and clear demonstration of the ability to produce usable knowledge are the first steps toward that worthy goal. If the social sciences can meet their obligations toward this partnership, I believe that protected area managers, from local district ranger to park superintendent to national chief to the IUCN leadership, will do likewise. For these managers well understand that the management of protected areas in the 21st century, now so close, is necessarily the management of people.

Chapter 8

Biosphere Reserves: A Comprehensive Approach

Bernd von Droste

Editor's Introduction:
Biosphere reserves are a form of "bioregional management," where a strictly protected area (the core zone of the biosphere reserve) is surrounded by various forms of land use that are designed to be compatible with the conservation objectives of the entire area. In attempting to build significant research, development, and education components into the biosphere reserve concept, UNESCO has been providing living examples of expanding partnerships for conservation. Bernd von Droste (Director, World Heritage Center, UNESCO), who has been involved in biosphere reserves for more than a decade, describes the experience of UNESCO to date, frankly assessing successes and failures. He calls for even more partnerships, including with business and industry, in order to address the needs of the "clients" of protected areas.

Introduction

Our horizons and potential for new partnerships have been significantly expanded over the last 10 years. Given my responsibilities for both UNESCO's Man and the Biosphere Program (which now involves 300 biosphere reserves in 76 countries) and the natural part of the World Heritage Convention (with 379 cultural and natural World Heritage Sites in 80 countries), I have been in the unique position of observing and helping many new partners work closer together. It has not always been an easy process, or a successful one, but we have learned many lessons along the way.

I recall briefly the well-known problems that we have already heard much about during the last decade.

- The need for transdisciplinary and interdisciplinary work, not just bringing together specialists in a multidisciplinary approach

- The obstacles of language, distance, and culture, which we must not simply try to overcome but rather exploit in the best sense of this term
- The old standby, lack of money

On the last point, I think we should acknowledge that the international conservation business, if I may call it that, is thriving. We have never had it so good in terms of the number of publications, films, television shows, meetings, conferences, and conventions dealing with our issues. The amount of money spent and the number of people working in this field has also increased, although perhaps not always by as much as we would have liked.

Harnessing Partnerships More Effectively

Why is it, then, that we have not been able to harness in a more effective way such expertise, resources, influence, and goodwill to reduce if not restore the damage to our natural and cultural heritage? Why is it that, in spite of having won many battles, we continue to lose ground, literally and figuratively, around the world?

I believe that we must pay much more attention to reaching out and taking some risks at both the personal and institutional levels. We must recognize that we cannot do it alone and that we will have to give a little in order to gain a lot. As recent events in the former Soviet Union show quite clearly, the gains may not come right away and they may not come to those who had the courage to give, but they will come if we make the right decisions and stick by them.

Let us take the example of reaching out at the interpersonal level. A case can be made that many of us, as natural scientists, have not been taught to work with other people, let alone develop productive relationships. Unlike other professions where human contact and teamwork are valued assets, we have often been trained to work with data or with species other than our own, and to compete rather than cooperate with our peers.

This problem is even worse at the institutional level, where many organizations have developed more or less spontaneously given the lack of international programs dealing with conservation and sustainable development. This has resulted in a certain overlap, which is now exacerbated by the increasing competition for responsibility or "market share" in this field.

UNESCO's program on Man and the Biosphere (MAB) has sought to be innovative in the partnership field, partly because of its institutional setting at what we like to say is the crossroads of education, science, culture, and communication. In fact, UNESCO's neutrality and its broad intergovern-mental mandate have enabled its programs to act as both a real partner and an "honest broker" with many different organizations and governments around the world. There is, of course, a down side to this, which is revealed by the lack of credit and sometimes even recognition given to UNESCO pro-grams, which tend to be more concerned about the propagation of ideas than about trademarks or patents.

This brings me to my next point, which is that we should all be changing or at least adapting our working methods to enable us to build more effective partnerships and alliances not only within international conservation circles but particularly outside. We must change our attitude from one of protecting circumscribed areas and various species from outside influences to working with such influences in protecting *our common future*, both within and outside protected areas.

Let me hasten to add that I am not bringing up the old debate between preservationists and conservationists, which has no place in the 21st century. As it turned out, both sides in this debate won various battles over the last hundred years but in the end we all lost because attention was diverted from the real problem of the unsustainable society that was developing almost everywhere else. Now practically everyone is suffering or will suffer to differ-ent degrees from the consumer society and human population that has grown out of control and is threatening not just protected areas but the entire world.

We not only need to work more closely with business and industry, we need to adopt more businesslike and democratic methods of getting things done. We should take a hard look at the way we run our affairs in order to improve our efficiency and productivity. In this day and age, when entire countries and multinational corporations are going through painful restructuring, we should not delude ourselves by thinking that we can continue to rely on the institutions and mechanisms of the 1960s and 1970s.

We must be much more flexible and responsive to changing needs and demands, not only of our governing or funding bodies, but of our clients, including the other species who share the earth with us and those people who will inherit it. This also means that we must be prepared to make some hard decisions about both institutional and personal responsibilities, and we must be prepared to live with them.

Similar analyses are needed to improve the way in which we involve local people and grassroots organizations in order to ensure that the democratic revolutions sweeping many parts of the world do not leave us behind. We must be more transparent in the way we make our decisions and the way we implement our programs. We can no longer pretend to be the only experts who know what needs to be done, let alone actually doing it.

Partnerships: The UNESCO Contribution

UNESCO is responding to these challenges in several ways, notably through revising the MAB program, establishing an advisory body for biosphere reserves, and exploring the possibility of improving the legal status and protection at the level of the international network, perhaps through a protocol in the new Convention on Biological Diversity.

Efforts are being made by the countries concerned, for example, to make their biosphere reserves reflect more accurately both their natural and social systems. They will thus have more of the irregular but natural boundaries of landforms or migratory patterns and will pay greater attention to socioeconomic factors. This will, of course, be much more difficult to manage, but it is one of the first steps toward sustainable development. We must realize that the battles for protected areas and the biological and sociocultural diversity they are helping to maintain cannot be won in the areas themselves or even on their boundaries. Only by extending and adapting the techniques of sustainable development demonstrated by biosphere reserves to the real world outside will there be any hope of saving not only the resources within them but the much greater stakes found beyond.

This is where biosphere reserves and other protected areas that combine research with sustainable development can play a crucial role in the coming decade "after Rio." They are ideally placed to make the best use of scientific information and traditional knowledge to promote sustainable development through the impressive range of ecosystems and societies that they represent.

There is, of course, a big difference between naming a site as a biosphere reserve and managing it as one because of the difficulties involved in implementing the concept of integrated sustainable development. Management is crucial for this process to blend into regional, and even national, development and planning so that biosphere reserves do not become islands of hope in a sea of despair, to be inundated by environmental refugees.

Biosphere reserves are fertile grounds for ecologically sustainable management and should be used not just for static conservation but as investment folios to regain local natural capital. By showing how to evaluate the real economic worth of biosphere reserves, including their environmental and social values, we would be providing policymakers with a more accurate tool for sustainable development decisions in the rest of the country.

This is a crucial step because the implementation of biosphere reserves depends on a national policy to encourage sustainable development. In this respect biosphere reserves can be regarded as microcosms of global problems, where it is becoming increasingly difficult to expect underdeveloped countries to bear the full costs of protecting the global environment.

Conclusion

Biosphere reserves are not just local or regional entities but have the potential of being a global network. This network can help to meet the increasing need for worldwide comparative data to monitor long-term changes in our environment and ways to manage them, and also can provide ground-truthing for remote sensing. That is why biosphere reserves are receiving particular attention in the joint UNESCO–IUBS–SCOPE program on taxonomic inventories and in UNEP's integrated terrestrial monitoring program. We are now working on site characteristics to help establish scientific international comparative research and monitoring. Biosphere reserve managers are also receiving training in cooperation with the World Heritage Fund, and we hope that the Global Environment Facility will provide support to our partners in less developed countries so that the biosphere reserve network can be fully exploited.

Part II

PARTNERSHIPS WITH MAJOR SECTORS

Forestry and Protected Areas:
A Natural Partnership

B.C.Y. Freezailah

Editor's Introduction:
In many parts of the world, protected areas are the responsibility of ministries or departments of forestry. This has often been an extremely useful partnership, enabling land from the forestry estate to be transferred to the protected areas estate with relatively little bureaucratic difficulty. At the same time, the partnership also has some inherent tensions, as most forestry departments are primarily production-oriented rather than conservation-oriented. That said, enlightened foresters also recognize that protected areas are an important part of land use, as part of the overall national land use pattern. Even within timber concessions, certain areas should be protected against exploitation, because they protect sources of seeds, water courses, or very steep slopes. Further, even managed forests can make important contributions to national strategies for conserving biological diversity. One of the most important institutional partners in seeking to apply sustainable forestry has been the International Tropical Timber Organization. B.C.Y. Freezailah, the executive director of the ITTO, points out that the major reason for the unsustainable use of forests is that mere renewability does not guarantee sustainability, for management is the key factor that will determine whether a renewable resource under exploitation is sustained, depleted, or, in some cases, enhanced. Failure to achieve sustainable use in forestry is generally the result of human and institutional failure rather than the natural limits of the resource base. It is clear that a healthy and productive forest estate is an essential partner to protected areas where extractive logging is neither permitted nor necessary.

Introduction

As the end of the current century draws near, the challenges facing those responsible for managing natural resources are becoming increasingly well defined as individual communities, nations, and the international community struggle to balance the need for economic development on the one hand with the need to protect biodiversity and ecological life support services of the natural environment on the other. The problems facing humanity and the problems facing the natural environment have generated considerable debate. They have also forced communities at the local, national, and international levels to reexamine the role of existing institutions that have been established to ensure the conservation and management of natural resources.

Both forestry and protected areas are undergoing critical reevaluation from both internal and external sources. The IVth World Congress on National Parks and Protected Areas had as its focus the theme of "Parks for Life" and the prospects for enhancing the role of protected areas in sustaining society. It is therefore useful to consider some of the challenges to both forestry and protected areas, and to examine ways that they might be better integrated with each other to meet the needs for environmental protection and the growing number of humans populating our planet.

I wish, at the outset, to reflect briefly on a number of issues including the nature and history of forestry and some of the similarities between forestry and protected area management, the current threats to the natural environment, the key role of protected areas in protecting biodiversity and contributing to maintaining ecological support services, and the limitations of protected areas in achieving these objectives. I will then set out what I think are the prospects for addressing the problems of forestry and protected areas in a complementary way through the processes of integrated land use planning and integrated natural resources management.

A Perspective on Forestry and Protected Area Management

Forestry belongs to those activities that have become the interface for conflicting community demands. Furthermore, the sorts of pressures foresters have felt, in some cases for centuries, are increasingly being felt by professionals in the field of protected area management when they are called on by the societies they serve to both protect the resources they manage and maximize the contribution of those resources to the economic development of that particular society.

The Oxford Dictionary defines forestry as the science or art of managing forests. This somewhat cryptic definition begs the question of managing forests for what objective. However, if one looks at both the history of forestry and the current diversity of roles being played by members of the forestry profession in different parts of the world, it is wise to extend the definition to that of "the science or art of managing forests for a variety of objectives which will be determined by the social, economic, political, and institutional framework afforded by any given culture at any particular time."

Such a broad definition of forestry would include protected area management where the protected area in question covers forest lands. In addition, the very definition of forestry as both a science and an art is in itself relevant to the management of protected areas. Modern forestry and protected area management both involve the application of scientific knowledge to decision making in resource management. However, both involve much more than just the application of scientific knowledge, as both must seek to resolve at times conflicting community priorities about the nature, manner, and timing of resource use within the prevailing social and political framework of the particular society in question.

Because timber production has historically been the dominant, or at least the most obvious, economic use of forest resources, many people equate forestry and the forestry profession with timber production, or even more narrowly with timber harvesting. However, the days when societies expect just timber management from their foresters or their forest management institutions are coming to a close in most societies, and foresters—like most professionals—are expected to respond to a much wider range of community interests and expectations.

The same can of course be said of protected area managers. The linkages between the two groups of professionals can be perceived as:

- Concern for both the conservation and the management of a renewable but easily degraded natural resource base

- A need for science-based decision making in an environment of much less than perfect knowledge or understanding

- The need to ensure high levels of output with low levels of risk to both natural values and resource capabilities

- The need to respond to an increasing diversity of often-conflicting community demands

Threats to Biodiversity

The biological capital of forest resources is under increasing threats across the globe, and nowhere is it more the case than in the world's tropical forests where deforestation and forest degradation are eroding the earth's biodiversity at an alarming rate. It is now widely known that the world's tropical rainforests are being rapidly and unsustainably exploited and that the areas they occupy are contracting markedly. The rate of loss itself is a major issue in rainforest conservation and management, for there is considerable uncertainty about the actual rate of deforestation. However, it is clear that substantial areas of forest are being destroyed or degraded. Furthermore, there is emerging evidence that with some 2.1 billion hectares of tropical forest remaining, the rate of clearing of both closed and open tropical forests may have accelerated from some 11.3 million ha per year in 1980 to as much as 17.1 million ha per year in 1990, or some 1.2 percent of tropical forests per year.

The impact of such historically high rates of forest loss is still the subject of considerable uncertainty and scientific speculation. However, current and projected rates of tropical deforestation suggest that as many as 150,000 species per year could become extinct. It has been suggested elsewhere that humankind is presiding over a biological debacle that may mean the demise of up to one-fifth of all known species of organisms in the foreseeable future. In addition, wider environmental impacts such as accelerated erosion and stream sedimentation have also been documented, and recent estimates suggest that tropical deforestation could account for from 7 percent to as much as 31 percent of current worldwide carbon dioxide emissions from all sources. Thus, with the enormity of the potential implications of rapid tropical deforestation, and with the scale of land use change that seems likely to occur over the next decade or so, there is some urgency for both research into and the development of operationally significant policies and management responses.

Beyond the geography of tropical deforestation, however, conservation managers must concern themselves with both the causes and the consequences of deforestation if effective responses are to be developed. Key issues of importance here include the loss of biological diversity and the sustainability of various forms of rainforest use.

Protected Areas and the Conservation of Biodiversity
The biological significance of the world's tropical rainforests is now well established, as is the case for a systematic network of protected areas. However, on a global scale, the long-term preservation of large areas of undisturbed rainforest will be difficult to achieve given the pressures for forest destruction resulting from population growth and the struggle for economic development in many of the countries with tropical rainforests. Nevertheless, options for the establishment of protected areas still exist in many areas. For example, it is estimated that as of 1990 there were still some 600 million ha of undisturbed but potentially exploitable closed forests and some 290 million ha that are currently left "unused" due to either legal protection or inaccessibility for exploitation. However, with the rate of forest conversion and degradation from poorly managed forest utilization, it is clear that systems of rapid resource appraisal are needed to help identify key and endangered sites so that appropriate systems of protected areas can be developed before irreversible land use changes occur.

The development and application of rigorous selection systems for protected areas offer the potential to blunt some of the projected losses in biodiversity in the wake of widespread tropical deforestation, because species distributions in tropical forests are not random. Thus, it has been suggested that there are a critical number of sites for species diversity and endemism which, if protected, can help to secure large proportions of the biodiversity stored in the tropical forests.

Equally, it is widely agreed that management of these key protected areas cannot be undertaken in isolation from that of surrounding lands subject to greater pressures from direct human utilization pressures. It has been noted that many tropical forest areas have expanded and contracted as climate has fluctuated in the past, and that their current natural extent in many areas is thought to be at or near the maximum achieved in the Quaternary. These secular changes in climate have left their imprint on the landscape and the geographical distribution of many plant and animal species, and the loss of retained buffers of managed forest to other nonforested land uses will reduce the ability of key protected areas to adapt to future changes in climate.

Many experts contend that very little tropical forest is pristine in the sense that it has never been inhabited by human beings. Large areas such as the Peten in Guatemala, Mexico, and Belize have been exploited for centuries, and what, in the latter case, appears to be a vast area of pristine forest is actually a mosaic of forest patches that have "recovered" from small-scale agriculture and forest clearance over many years in times past.

We need to understand this fact clearly if we are to conserve the biological diversity of tropical forests in the face of the seemingly devastating rates of forest conversion and forest degradation that are occurring today. Thus, if protected areas and the management of natural forests are to be included as an important part of the global solution to deforestation, land use planning and zoning will be needed on a scale vastly greater than currently practiced in the tropics. In this context, forest management must be considered in the broadest sense of land use planning, including clearing the forests for shifting cultivation, habitation, and hunting; the voluntary and intentional protection of forests or individual species; the extraction at various levels of intensity of a wide variety of forest products; and the enrichment of the forest with species from other places. Forest management also needs to be considered in terms of the inhabitants of these regions, whose knowledge of the trees and the forest is also a largely untapped resource that is likely to be lost before it can be studied or preserved.

Achieving this broad approach to the conservation of tropical forest resources will require considerable support. UNEP estimates that the successful implementation of the Convention on Biological Diversity would require funding of between $1 billion to $10 billion per year for the next 10 to 15 years. However, in the context of rapid land use change outlined previously, the need for immediate action is self-evident. It is also clear that there is a considerable need for the transfer of knowledge and technology from the developed economies to the developing tropical countries and, just as important, between the tropical countries themselves.

Sustainability and Integrated Land Use Management

In 1980, sustainable utilization of species and ecosystems was a key component of the overall sustainable development strategies embodied in the *World Conservation Strategy* (WCS). This concept has been reinforced in subsequent documents, most recently in the update of the WCS, *Caring for the Earth*. It is also increasingly being seen as an important component of strategies to promote the retention of extensive areas of tropical forest for the conservation of biological diversity itself.

The economic imperative behind the search for sustainable utilization systems is twofold. First, the economic value of the forest-based industries is of considerable importance to a number of tropical countries from the perspective of both employment and foreign exchange earnings. While some countries appear to be deliberately liquidating their forest resources to generate capital to initiate the development of industrial infrastructures in much the same way as some European countries did at the start of the industrial revolution, many will face significant economic and social hardships if the substantial forest industries that they have put in place cannot be sustained and alternative sources of income and employment are not developed.

Second, patterns of land use need to be found within which it can be demonstrated that the retention of substantial areas of natural and near-natural tropical forest is in the interest of developing countries and the local people living in and near the forest as well as to the global community as a whole. In short, mechanisms must be found so that developing countries and local communities gain more from keeping their forests intact than from clearing them for nonforest development activities.

Part of the answer to this problem is for the developed economies to pay developing countries directly for the global environmental services developing countries produce by retaining intact tropical forest ecosystems. However, in the absence of the extensive funding outlined previously, and with the reluctance of some countries of the South to become what they see as mere custodians of extensive botanical gardens for the countries of the North in much the same way that they once became the breadbaskets for the former colonial empires, sustainable management becomes a key conservation goal. In its absence, the maxim "use them or lose them" will remain true for many rainforest areas and will continue to be reflected in the prevailing high rates of tropical deforestation.

Sustainability is a concept that is simple in principle but very difficult to achieve in resource management practice. In theory, it should be possible for human beings to utilize indefinitely any living resource that is characterized by cycles of birth, growth, death, and recycling by matching the intensity of their utilization patterns to the long-term growth potential of the living resource being exploited. However, despite this apparent simplicity, there are very few examples in either the temperate or the tropical worlds where major potentially renewable resources such as forests or fisheries have been managed sustainably over an extended period of time.

The major reason for this situation is that mere renewability does not guarantee sustainability. Management is the key factor that will determine whether the productivity of a renewable resource under exploitation is sustained, depleted, or, in some cases, enhanced. Thus, failure to achieve sustainability in natural resource management is generally the result of human and institutional failure rather than due to the inherent sustainability of the resource itself.

Scholars have reviewed the application of sustained yield concepts to tropical forest management in northeastern Australia and elsewhere and have made a number of important points about the classical forestry paradigm of sustained yield and the more modern concept of sustainability. They note that the sustained yield paradigm essentially evolved in Germany in the mid-19th century after German forests had experienced a long history of deliberate manipulation and management and that, as a result, old-growth trees were essentially absent from those forests. In essence, when areas had reached a desired degree of maturity, they were cut, the forest was regenerated, and the new generation of trees was grown through the new cycle to be ready for harvesting again at some future date. In addition, by having equal volumes of timber ultimately produced from each forest age class from the so-called normal or regulated forest, the harvest each year or at each interval could be approximately the same.

While these conditions applied in Germany and to some extent in other European forests, they did not apply to many forests in North America, Australia, and the tropics due to the dominant presence of old-growth forest stands. In these forests, the initial timber management problem was to oversee the orderly—or in many cases not so orderly—liquidation of the accumulated timber volumes during an extended transition period that would ultimately lead to a yield allocation that could be sustained by the growth increment of the managed forest alone. Thus, some experts contend that there is no known example in the world of sustainable logging in tropical rainforests that would take about three cutting cycles, about 150 years, to demonstrate. This is also true for the vast majority of the world's temperate forests. As a study for the UNCED (U.N. Conference on Environmental Development) has noted, detailed forest harvest regulation and working plans have been common in most of Europe only since the 19th century. In the boreal zone, the Scandinavian countries have had only several decades of experience with the sustainable management of productive forests through intensive management. Sustainable management has been achieved to a lesser degree in Canada and in the boreal forests of the former U.S.S.R.

While with time concepts of sustained yield in some forests had become more focused on ideas of nondeclining yields, more important legislative changes in the United States in the 1960s had broadened the concept to include the conservation of nontimber forest values and ecological concepts such as maintaining site productivity. Thus, the more modern sustained yield forestry concepts tend to merge with contemporary ideas about resource sustainability in general and point to the idea of achieving and maintaining in perpetuity high-level annual or regular periodic outputs of the various renewable resources without impairment of the productivity of the land.

Recent statements on the sustainable management of tropical forest lands outline similar concerns, suggesting that timber production from natural tropical forests should be considered sustainable only in a land use planning context that allocates forest lands for the total protection of biodiversity and fragile environments such as steep slopes and critical watershed areas. Clear management objectives need to be specified for each tract of forest to ensure an overall balance of uses. Reference is also made to issues such as operational control, research, and training as essential elements of sustainable production.

The issue of the balance of uses is particularly important because it is increasingly being recognized that tropical forests have been economically undervalued in general and that the value of nonwood forest products and environmental services have been economically undervalued in particular. As an example of the latter, it has been noted that in the western Amazon, annual incomes of up to US$ 200/ha could be produced from the nonwood forest products, such as, nuts, fruits, wildlife, and medicinal plants, if they were harvested on a sustainable basis.

In terms of the integration of forest conservation with the production of economic benefits for local people, this latter finding is quite significant because nonwood forest products can often be harvested with little or no disturbance to the forest soil or the standing vegetation. In addition, many nonwood forest products can be harvested annually, which gives them a considerable economic cash flow advantage over the long time intervals between timber harvests. Thus, while a specific timber harvest will often have much greater monetary value than the harvesting of nonwood products, the net present value of harvesting nonwood products over time may outstrip those of timber harvesting. However, on the debit side, it must be cautioned that the continuous harvesting of some nonwood products may place larger long-term nutrient demands on the site than irregularly spaced timber harvests and therefore may not be as easily sustained ecologically or economically on marginal sites.

Conclusion

Thus, while protected areas have traditionally been the major tool for land managers to promote biological conservation, the lessons of island biogeography and conservation biology tell us that the network of totally protected reserves we have established through such protected areas are never by themselves adequate to conserve all, or even most, of current existing biodiversity. Integrated management of the whole landscape, including forests allocated for timber production, is now both a social and an ecological necessity.

National parks and other totally protected areas must of course still be seen as the jewels in the crown of any nation's endowment of natural resources. However, with the degree of land use change and population pressure in most of the tropical world, other forest lands must also provide much of the substance of that crown. They must therefore be managed accordingly to provide both immediate economic benefits for the communities that live in and near the forests, and the broader range of environmental services that the global community now recognizes as being essential for the future of humankind.

In particular, the production forests of the tropical world must be managed as a vital complement to our totally protected forest reserves, so that extensive areas of natural and near-natural forest are retained to secure as full a complement of biodiversity and options for future generations as is humanly possible.

In the final analysis we hold it as a supreme proposition of action that international cooperation can only be useful, valid, and salient if it is practical—not theoretical; if it has a human face and dimension in the sense of its promoting the day-to-day needs for human livelihood; and if it receives full financial and material support through solidarity of the world community. The world community, in its eagerness to ensure effective management of the world's natural resources, including forests and protected areas, should remember that its eagerness and concerns must necessarily be translated into a sense of urgency and responsibility to mobilize actions of effective support and solidarity with the forest communities concerned if the world strategy on global sustainable development is ever to be realized. Anything short of this act of political will and solidarity will be an exercise in futility.

Chapter 10

Medicinal Plants and Protected Areas

Olayiwola Akerele

Editor's Introduction:
As the rising tide of humanity reduces the extent of natural vegetation, those who are interested in medicinal plants will need to look increasingly to legally protected areas as the last reservoir of plants and animals with enormous genetic significance. Equally, those who are charged with managing these areas will need to accommodate their management approaches so that sustainable supplies of medicinal plants and animals can be provided. As the manager for the Traditional Medicine Program of the World Health Organization, Dr. Olayiwola Akerele calls for a new partnership to be formed between private industry, those institutions involved in medicinal plants (including ministries of health, indigenous peoples, and rural development as well as universities, botanic gardens, and research institutions), and those institutions involved in conservation of biological diversity (including protected area management authorities, donor agencies, and nongovernmental organizations). The major international development institutions, such as FAO, WHO, UNDP, and the World Bank, also need to support this linkage between effective management of protected areas and the provision of sustainable benefits such as medicinal plants to human societies. At the national level, ministries of planning, finance, and rural development need to be made aware of this new partnership, and need to take steps to foster more productive collaboration.

Introduction

The discovery and use of sulfa drugs, as well as antibiotics such as penicillin, led to a dramatic decline in the popularity of medicinal plant use in therapy. Now the pendulum has swung in the other direction. A resurgence of interest in the study of medicinal plants has been taking place during the past two decades. Popular, official, and commercial interest in the use of natural products has also grown considerably. The World Health Organization (WHO) has promoted awareness of the importance of traditional medicine to the

75

majority of the world's population and has promoted rational exploitation of safe and effective practices, including the use of medicinal plants.

Much, if not most, of what constitutes modern drugs owes its presence directly or indirectly to chemicals originally found in plants. In spite of this, the vast majority of plants have never been assessed pharmacologically for potential medicinal value, even those that are currently being used for medicinal purposes by indigenous people. Thus plants represent a largely untapped resource but, unfortunately, one whose very existence is being threatened by the same forces that are contributing to the destruction of the world's great forests. It has been estimated that no less than 60,000 plants, nearly one in four of the world's total, could become extinct by the middle of the next century, if present trends continue. The concern with medicinal plants therefore is intimately tied up with the conservation of nature in all aspects, including protected areas.

Most of the plant variety in the world is to be found in the tropics and subtropics, which are the least developed areas in social and economic terms and most in need of more effective and lower-cost healthcare. Thus the full exploitation of medicinal plants must be seen as part of the larger need for national socioeconomic development, the improvement of healthcare systems, and the preservation of a natural resource base for sustainable use in the future. It would, therefore, not be an exaggeration to say that what we do to meet the genuine aspirations and needs of the developing world is an important contribution to other measures intimately linked with the preservation of our global village.

This chapter will focus on the socioeconomic importance of medicinal plants and how protected areas can collaborate most productively with the health services system. WHO's program in traditional medicine touches on all aspects of its work in promoting the use of medicinal plants. While progress has accelerated in recent years, as long as there remain any plants with medicinal promise that have not been examined for potential use as a therapeutic agent, and whose conservation is not assured, we should not be content with our progress. As only a small percentage of plants have been explored and protected in this way, it is clear that our work has only just begun. Nevertheless, progress to date has helped to clarify where our priorities lie and what lines of action to follow. The task now is to get on with mobilizing the necessary financial and human resources to see that implementation follows the policy directions that have been agreed upon.

The Traditional Medicine Program of the World Health Organization is based on the reality that (1) the majority of the world's population depends on traditional medicine for primary healthcare, (2) the manpower represented by practitioners of traditional medicine is a potentially important resource for the delivery of healthcare, and (3) medicinal plants are of great importance to the health of individuals and communities. Collaboration with protected areas will be based on these understandings.

In March 1988 an International Consultation on the Conservation of Medicinal Plants was convened in Chiangmai, Thailand, by WHO in association with IUCN and WWF. One outcome of this meeting was the adoption of the Chiangmai Declaration entitled "Saving Lives by Saving Plants" (Akerele, Heywood, and Synge, 1991).

It is worth emphasizing that the drive toward "Saving Lives by Saving Plants" is a part of WHO's wider concern for the environment. Medicinal plants are a small but important part of the biological heritage of the earth. Traditional societies, also rapidly vanishing, place a high value on this inheritance, which they express through their intimate relationship with and respect for nature. What is significant today is the growing recognition from individuals and governments of the industrialized world that these so-called traditional values are valid for all people. Responding to the environmental and ecological deterioration that threatens development everywhere, a worldwide movement has arisen to awaken humanity to the dangers facing our planet and to help preserve its ecological integrity.

Thus, policies regarding traditional medicine, especially as they relate to medicinal plants, have evolved over the last 10 to 15 years from a somewhat passive recognition of their roles in the provision of medical care to an active promotion of their identification, conservation, and further exploitation, especially in the light of the alarming loss of plant diversity around the world due to habitat destruction and unsustainable harvesting practices.

Importance of Plant-Derived Drugs

The success of any health system depends on the ready availability and use of suitable drugs on a sustainable basis. Medicinal plants have always played a prominent role in healthcare, and plant-derived medicines are used in self-medication in all cultures. It is within that context that WHO considers the important role that plants can play in the health of humankind.

Medicinal plants are commonly available in abundance, especially in the tropics. They offer the local population immediate access to safe and effective products for use in the prevention and treatment of illness through self-medication. Furthermore, they are valuable for modern medicine in four basic ways (1) they are used as sources of direct therapeutic agents, (2) they serve as a raw material base for the elaboration of more complex semisynthetic chemical compounds, (3) the chemical structures derived from plant substances can be used as models for new synthetic compounds, and (4) plants can be used as taxonomic markers for the discovery of new compounds.

Many examples of the benefits that have already been drawn from tropical plants can be provided. A number of essential drugs used in modern-day medical practice are plant-derived: atropine (anticholinergic), codeine (antitussic/analgesic), colchicine (antigout), digitoxin/digoxin (cardiotonic), vincristine (antitumor), morphine (analgesic), quinine/artemisinin (antimalarial), reserpine (antihypertensive), and physostigmine (cholinergic). Many more well-known examples can be cited.

For the developing world, the use of medicinal plants based on local production also offers the possibility of reducing dependence on imported drugs requiring the expenditure of scarce hard currency, and at the same time promoting economic self-reliance and cultural acceptability of the medicine(s) offered. The government of Costa Rica, for example, recognizes the importance of plants, including medicinal ones, and has allocated 25 percent of its land as forest preserves. The National Institute of Biodiversity of Costa Rica (INBio) has signed an agreement with the US firm Merck and Company by which INBio will provide Merck with plant samples as well as the opportunity to evaluate these samples for pharmaceutical and agricultural applications. In return, and as part of the two-year agreement, Merck will provide INBio research funding annually plus certain start-up expenses, as well as royalties on the sale of any products that Merck ultimately develops from an INBio sample. Part of the research funding will pass directly to the Costa Rican National System of Protected Areas for conservation of biological diversity in the field from which these samples are drawn.

WHO has estimated that perhaps 80 percent of the more than 5000 million inhabitants of the world rely chiefly on traditional medicines for their primary healthcare needs, and it can safely be presumed that a major part of traditional therapy involves the use of plant extracts or their active principles.

A computerized database on medicinal plants at the University of Chicago has documented ethnomedical uses alone for about 9200 of 33,000 (ca 28 percent) species of monocots, dicots, gymnosperms, pteridophytes, bryophytes, and lichens. In the People's Republic of China, it is reported that 5000 of 35,000 (ca 14 percent) species of plants still growing are used as drugs in Chinese traditional medicine. Taking the range of 14 to 28 percent as being an indication of the percentage of plants that are used as drugs, and accepting the conservative number of 250,000 higher plant species on earth, one can estimate that from 35,000 to 70,000 species have at one time or another been used in one culture or another for medicinal purposes (Farnsworth and Soejarto, 1991).

Turning to the economic value of medicinal plants, data are available from only a few countries. For example, the annual production of traditional plant remedies in China was valued at US$ 571 million, and countrywide sales of crude plant drugs at US$ 1400 million in 1985 (Akerele, Stott, and Lu, 1987). In the United States, 25 percent of all prescriptions dispensed from community pharmacies from 1959 to 1980 contained plant extracts or active principles prepared from higher plants, estimated in 1980 to have cost consumers more than $8 billion (Farnsworth and Soejarto, 1985).

Protected Areas and the Conservation of Medicinal Plants

The Chiangmai Declaration brought together several major constituencies that until then were working largely independently of each other, namely health authorities concerned with the vital importance of medicinal plants in healthcare, biologists concerned with the loss of biodiversity, and environmentalists concerned with the rapid deterioration of the world's natural resources. Added to this mixture of concerns there was also the view that medicinal plants offer a valuable resource for developing countries, which both scientifically and medically had not begun to be tapped due to the lack of adequate scientific, technical, and commercial infrastructures. In other words, not only will saving plants lead to the saving of lives, it will lead to future scientific and economic returns of great importance to indigenous people where the plants are to be found, and to the rest of the world. But for this future to be realizable we must conserve what nature has given us.

Broadly speaking the strategy for conservation is developing along two major lines: the global necessity to prevent the disappearance of forests and associated biological species, including plant life; and the local necessity to establish botanical gardens with strengthened capacities to achieve conservation. IUCN and WWF, the world's largest independent conservation organizations, have led the way in outlining the details of both strategies as they apply to plants in general and more recently to their medicinal component. They have increased public awareness of how plants contribute to our lives, and have pressured governments and industry to develop and carry out conservation policies.

WHO is fully mandated by its policymaking bodies to form partnerships with the conservation movement to ensure that adequate quantities of medicinal plants are available for future generations. With respect to the global concern with disappearing natural resources, WHO is participating actively in international efforts toward sustainable development. Much remains to be done in this area. Few developing countries, for example, have instituted measures to protect the endangered species of medicinal plants. Most medical schools and faculties have no place in their curriculum for botany and no access to botanical gardens where the study of the role of medicinal plants might be pursued. Furthermore, out of the total of more than 1400 botanical gardens in the world, only some 230 are to be found in the tropics and subtropics, and most of these are in Asia. Latin America and tropical Africa are highly underdeveloped in this regard. These are elements of a long-term strategy that must receive greater attention if the marriage of traditional and modern medicine is to take firm root.

The Chiangmai Consultation outlines the role of botany in national programs concerning medicinal plants, but it did not specify the key functions that can be played in such programs by botanical gardens, national parks, and protected areas in general. McNeely and Thorsell (1991) noted that there is virtually no long-term conservation experience dealing specifically with medicinal plants in protected areas. They outlined some useful and practical guidelines for the selection, planning, and management of medicinal plants in protected areas. Among other aspects, they suggested a review of what is known about the distribution of original and existing vegetation in the country, identification of medicinal uses of plants, their distribution and abundance, an assessment of the extent to which the existing protected area system incorporates the plant diversity of the nation, and preparation of management plans for utilizing the medicinal plants on a sustainable basis.

Conclusion

The proper use of medicinal plants is a necessity, not a luxury. A critical element for the success of any healthcare system is the availability and use of appropriate drugs, including those from medicinal plants.

The challenge of conserving medicinal plants—and developing the role of protected areas in doing so—is particularly acute, as the potential of activities in this area is at present far in excess of the resources readily available either to WHO or, more important, to the developing countries where the need is greatest. Clearly a major problem remains how to mobilize more resources, while at the same time remaining faithful to the desire that activities carried out should first and foremost benefit the developing world. Many WHO programs face the dilemma that the scientific and technical know-how is almost all to be found in developed countries. This dilemma applies to the subject of medicinal plants, where little local capacity is to be found, with the exception of a handful of countries and the countless traditional practitioners who demonstrate their expertise in their daily work. If we add the commercial dimension, which also has its economic and human demands, the dilemma is further accentuated.

Paradoxically, it is the commercial factor that probably provides the incentive and a key to developing a strategy that could best safeguard the interests of the developing world. But such a strategy can only succeed against the background of a responsible public strategy, one that fully recognizes the importance of the conservation of natural resources and supports the sustainable exploitation of such resources for the overall public good. It is this balance of interests that forms the essential base of activities in this exciting and challenging area.

References

Akerele, O., V. Heywood, and H. Synge (eds.). 1991. *Conservation of Medicinal Plants.* Cambridge, United Kingdom: Cambridge University Press. 199–212.

Akerele, O., G. Stott, and Lu Weibo (eds.). 1987. The role of traditional medicine in primary health care in China. *American Journal of Chinese Medicine*, Suppl. No. 1.

Farnsworth, N.R., and D.D. Soejarto. 1985. Potential consequence of plant extinction in the United States on the current and future availability of prescription drugs. *Economic Botany* 39: 231–241.

———. 1991. Global importance of medicinal plants. In *Conservation of Medicinal Plants* O. Akerele, V. Heywood, and H. Synge, eds. Cambridge, United Kingdom: Cambridge University Press. 25–51.

McNeely, J.A., and J.W. Thorsell. 1991. Enhancing the role of protected areas in conserving medicinal plants. In *Conservation of Medicinal Plants*. O. Akerele, V. Heywood, and H. Synge (eds.). Cambridge, United Kingdom: Cambridge University Press. 13–22

The Evolution of Zoos, Aquaria, and Botanic Gardens in Relation to Protected Areas

George B. Rabb

Editor's Introduction:
For most people, zoos are at the opposite end of the spectrum from protected areas. But that relationship is changing, as this chapter, prepared by the director of a prominent zoo and the chair of IUCN's Species Survival Commission, demonstrates. In fact, zoos, aquaria, and botanic gardens are a natural part of the development of a conservation ethic among the general public. These are places where people can go to see the animals that are often reclusive, even in protected areas, and to build feelings of respect for these creatures. Further, as this chapter demonstrates, much of the management science that will be required to restore degraded habitats and to manage small populations of wildlife are being developed in zoos and botanic gardens. These so-called *ex situ* facilities are also playing an increasingly important role in generating public support, and especially finance, for protected areas. In addition, growing numbers of protected areas have formal links with adjacent or nearby *ex situ* facilities, providing a means for delivering additional benefits of protected areas to the public.

Introduction

Given the increasing pressures of the human species on the rest of the world, literally crowding out nature, it is necessary to employ a range of strategic responses if we are to have and leave to the future many parts of the global biota. These responses extend from seed and animal genome banks to designated wilderness areas. In between as conservation areas are zoos, botanic gardens, and aquaria as *ex situ* representations of diverse biotic elements, and national parks and similar protected areas as specially delimited natural habitats. Observe that both the zoos and related institutions as well as protected areas are organized by people—although the parks contain natural things, they are nevertheless human constructs. They are therefore susceptible to change by us.

In the case of zoos, progressive change can be seen as an evolution in their missions over the last century and a half. Figure 11-1 illustrates the changes in concept, the biological focus, the themes offered to the public, the institutional concerns at different stages, and the mode of display or communication to the users or visitors. Although modern zoos are now very much engaged in managing their collections for long-term survival of species and are trying to practice conservation consistently in all their operations, they still perform valuable functions that are attributes of earlier stages in their development. More particularly, it is still important for zoos to illustrate the diversity of the animal kingdom, and to inform people about the ecological relationships among species of plants and animals and the behaviors that enable animals to exist as populations in their natural habitats. Nevertheless, these institutions can be of even greater value by continuing to move to a role as centers for the conservation of biological diversity.

A remarkable parallel to the evolutionary path for zoos can be seen in regard to the management of national parks and other protected areas. About a century ago, national parks were conceived as places of exceptional natural beauty or quality that should be allocated for the enjoyment of people as visitors. The concept has changed to embrace biosphere reserves, with envelopes of increasing human habitation and activity. However, even more change is evident in the biologically informed management that is necessary if protected areas are to maintain their ecological integrity and conserve biotic diversity. In particular, since these areas are limited and migrations are no longer possible, large animal populations have to be kept within the ecological carrying capacity of the environment. In national parks like Kruger in South Africa, the park management provides for the water supply, periodic renewal of vegetation for food, monitoring of wildlife diseases, and regulation of the numbers of major mammal species to keep within the carrying capacity of the park. Such protected areas are megazoos in essence, for these are precisely the basic functions that zoos discharge to maintain their animal inhabitants. Thus these two kinds of facilities have come to have much in common with respect to the fundamental concern for maintenance of populations of animals. And both are now very much involved in long-term conservation measures for those populations.

Menagerie

Living Natural History Cabinet

Theme: Taxonomic

Subjects: Diversity of species
Adaptations for life

Concerns: Species husbandry
Species propagation

Exhibitry: Cages

19th century

Zoological Park

Living Museum

Theme: Ecological

Subjects: Habitats of animals
Behavioral biology

Concerns: Cooperative species management
Professional development

Exhibitry: Dioramas

20th century

Conservation Center

Environmental Resource Center

Theme: Environmental

Subjects: Ecosystems
Survival of species

Concerns: Holistic conservation
Organizational networks

Exhibitry: Immersion exhibits

21st century

Cooperation between *In Situ* and *Ex Situ* Forms of Conservation

Possible areas of cooperation between zoos, botanic gardens, aquaria, and other *ex situ* forms of conservation on the one hand, and protected areas as a major form of *in situ* conservation on the other hand, can be summarized as follows:

1.   The primary service of zoos to their visitors is in education, or communications about the animals and the environment. Clearly zoos could be embassies for national parks, just as the animals in zoos have come to be considered ambassadors for their species. Thus far the former has been done as matter of course in only a few zoos. Zoos in developed countries have specific opportunities to communicate about the conservation roles of protected areas as they host visiting professional staff from national parks of other countries.

2.   Zoos can serve as training places for qualified students and researchers to learn about animals before undertaking field research in protected areas. This makes for more efficient and effective use of time in the field. Zoos can also help by training people involved with protected areas in techniques and disciplines that may be needed for study or management in the field (such skills as immobilization of various classes of animals, behavioral indices of social group dynamics, genetic sampling requirements, and so on). Such short-term training has been taken outside the zoos through a cooperative international program coordinated by the U.S. National Zoological Park. The Jersey Wildlife Preservation Trust in the United Kingdom has sponsored resident training programs of this kind for some years.

3.   Zoos increasingly serve as refuges for species threatened with extinction in protected areas. This has been true for a long time (the American and European species of bison are examples from early in this century), but has been adopted as a principal way for zoos to contribute to wildlife conservation over the last two decades. More than 100 collective species-survival management plans are now in place in zoos around the world. The stocks of species so managed in zoos and allied facilities are now being used for reintroductions to areas of natural or restored habitat (the golden lion tamarin of eastern Brazil is a recent example).

4.   Some zoos are engaged in direct outreach—financing in part the basic maintenance of a national park. The Minnesota Zoological Park has adopted the Ujung Kulon National Park in Indonesia, for example, extending the zoo's general interest in the fauna of southeastern Asia to this action on behalf of the highly endangered Javan rhinoceros. Zoos can indirectly pro-

mote the welfare of national parks by sponsoring tours by their members and visitors. This has become a normal part of operations in many North American and some European zoos.

5.   Field research, monitoring, and other work on protected areas is a widespread service of zoos and botanic gardens. There are about 250 instances where botanic gardens are so involved, often as part of a national park, and perhaps 150 cases of zoos providing such collaborative help. Notable is the long-term effort of the Frankfurt Zoological Society in the protected areas of Tanzania and Zaire. Unquestionably the most significant program worldwide is that of Wildlife Conservation International, a branch of the New York Zoological Society, which has been involved in 90 such projects in 30 countries.

6.   The collaboration of zoos through the Captive Breeding Specialist Group of SSC/IUCN has led to several services of direct relation to the conservation interests of protected areas. The application of small-population biology principles to the management needs for wild populations is a recent development that is being spearheaded by this specialist group. Population viability analyses and risk analysis utilizing the experience of the zoo community and the expertise in the SSC network of taxonomic specialist groups are proving to be powerful tools in organizing information for conservation of species in protected areas. An outgrowth of such work has been the formation of "faunal interest groups" to guide the involvement of zoos in conservation of the faunas of various countries. Such groups now exist for Madagascar, Indonesia, Central America, and Zaire. They are engaged in helping educational, marketing, captive husbandry, and monitoring efforts of local zoos as well as national parks.

Expanding Cooperation in the Future

In looking at the conservation engagements of zoos in the developed countries, some general recommendations for successful partnerships of governments, protected area management authorities, and zoos emerge.

- There should be commonality of interest. That is, a zoo should have some reference point in its collection or exhibits or research programs that makes relevant and understandable its involvement with protected areas or *in situ* programs. It is otherwise difficult to sustain such an effort beyond the initiating persons.

- The commitment should be for a long-term, multiyear relationship on both sides. The expense and energy devoted to short-term endeavors are scarcely warranted, with the exception of emergency rescue efforts for highly endangered species populations. Even the latter have potential for misunderstanding about motivation and therefore should have clearly stated objectives and terms.

- Projects and programs of modest costs and visible benefits are most desirable because of the limited financial capacities of most zoos and of the likelihood that the contributors will desire an accounting of tangible products. However, the potential for fund-raising at modest levels is considerable, since it can be spread over many communities.

- A marketing and communications program should be built into cooperative relationships to maximize the effects of informing and educating people in the local communities and elsewhere (both the home territory of the zoo and of the protected area). This could extend to fund-raising among foundations and aid agencies and to the recruitment of volunteer efforts in the parks from the zoo's audience or membership.

Conclusion

From the above it can be appreciated that zoos and protected areas can associate in many ways to mutual benefit in their wildlife conservation functions. We must do a better job in ensuring that these facilities and the biota they care for endure. In this connection, protected areas generally need much enhanced communication and marketing to their publics and governments if they are to compete successfully for ongoing national support. Protected areas might therefore well consider the desirability of a strategy of "adopting" zoos in their own countries in order to reach the urban populations who will in one manner or another substantially influence the existence and continuation of protected areas or other managed ecosystems. Such a strategy could also benefit the collaborating zoos by offering them a heightened purpose and in the process provide clear targets for their exhibition programs—immersing visitors in simulated or actual environments similar to those of the national park partners. Indeed, if protected areas are to become exemplars of how to manage natural environments, rather than island fortresses for wildlife protection, and if zoos, aquaria, and botanic gardens are to maintain relevancy to human society, then such evolutionary steps in relationships is not an option—it is an obligation.

Chapter 12

Fisheries and Protected Areas

James M. Kapetsky and Devin M. Bartley

Editor's Introduction:
In many parts of the world, fisheries are the most important source of protein; they also provide significant employment. This chapter, written by James M. Kapetsky and Devin M. Bartley, two fisheries experts from FAO's Inland Water Resources and Aquaculture Service, suggest what can be done to promote multipurpose protected areas that can sustain fisheries as well as promote the conservation of other living resources. Protected areas are able to contribute to fisheries through conserving critical habitats, maintaining habitat continuity, and supporting the maintenance of essential ecological processes. Important habitats for fisheries are often highly dynamic, requiring seasonal inundation or periodic openings to the sea. Maintaining both habitats and their full range of functions are essential to sustaining their contributions to fisheries. This may require protected area managers to take a somewhat more dynamic view of boundaries and various resource management issues, and to form partnerships with fisheries agencies.

Introduction

An increasing need to protect fishery resources by maintaining essential processes and by conserving the critical habitats and systems on which they depend is ample incentive for fisheries agencies to promote protected areas in cooperation with other organizations. Agencies can form partnerships for conservation; however, protected areas initiated for other purposes will have the best chances for success if they incorporate fisheries objectives where fisheries are already important. To make protected areas work, the task is to promote communication that identifies where one agency's mandated responsibilities and interests coincide with those of other organizations.

In the context of this chapter, we choose to assign a utilitarian motive and function to aquatic reserves. That is, the primary goal will be to increase the quality of the human condition. We recognize the wide variety of justifications for protected areas, ranging from preserving the aesthetic value of natural features to managing living aquatic resources for sustainable fisheries among the poorest countries. But coastal areas are popular centers for human population development, and river systems have historically been avenues of transportation and commerce. Therefore, the majority of aquatic reserves must incorporate anthropocentric values and concerns; preservation and indeed rehabilitation of vast expanses of pristine coastline or of river basins may be unrealistic.

There are two basic kinds of protected area–fishery relationships.

• *Protected areas for fishery management*, in which protected areas figure as spatial management tools to maintain or increase benefits from fisheries. The protection for fisheries purposes may produce side benefits for other uses, such as conservation of biodiversity and maintenance of genetic resources.

• *Fisheries in areas that are protected for other purposes*. Where the original fisheries were prohibited, or were at a low level of development before the protected area came into being, fisheries increase the overall value of the protected area by utilizing its fishery resources in a sustainable way. Where existing fisheries are already well developed, establishing a protected area may be advantageous for fisheries by providing nurseries or refuges. The fundamental problem is to be able to show conclusively that the long-term gains in fisheries benefits, such as increased yields, make up for the short-term negative impacts such as partial loss of fishing grounds.

These situations are elaborated below.

Protected Areas as Fishery Management Tools

Areas in which fishing is temporally or spatially regulated (that is, in which either resources or habitats are protected) have long constituted fishery management options for inland, coastal, and high seas fisheries. The "management" ranges from local and traditional control over access, such as by chiefs or priests, to increasingly higher levels of administration through international fishing conventions.

Three levels of protection in the fisheries context have been identified by Odum (1984). These are (a) protection of the organism while located in a critical habitat, (b) protection of specific habitats, and (c) protection of complete sets of critical habitats. The objectives of such protection are increased yield from fishing and/or value of the fisheries by producing larger, more desirable individuals or a greater diversity of fishes. These benefits accrue through control over fishing mortality, manipulating survival and growth for a better coupling of the biological productivity of a species with size-related harvest and market demand.

Spatially protected areas can be allocated for spawning grounds, nurseries, and spawning migrations. Usually these are aimed at one or a few related species, and the protection is seasonal rather than permanent. Caddy (1990a) documents significant yield increases in Mediterranean trawl fisheries for which seasonal and area closures have been adopted. A familiar example is the closure of certain portions of a river during salmon migration and spawning. In contrast to seasonal closures, marine fishery reserves may be permanently closed to fishing. The concept of marine fishery reserves for reef fishery management has been studied in detail (Plan Development Team, 1990) and is now receiving increasing attention (Biro, 1992). Specific benefits include protection of critical spawning stock, intraspecific diversity, population age structure, recruitment supply, and ecosystem balance. These benefits have long-term implications relating to sustainability of fisheries and to the genetic well-being of the stocks of fishes under exploitation.

Conservation of Biodiversity and Maintenance of Genetic Resources through Protected Areas

Aquatic protected areas can be effective tools to assist with the conservation of genetic resources, biodiversity, and ecosystems for the ultimate benefit of nature and humankind (Reid and Miller, 1989). In addition, aquatic reserves may be a means for effective management of fisheries when traditional catch and effort management strategies are ineffective or impractical (Roberts and Polunin, 1991). This is because the reserve provides an area with no or limited harvest, but it can supply recruitment to or refuge from adjacent areas that are fished.

Fishing pressures may result in changes in life history characters, reduction in age at maturity, or reduction in amount of genetic variation present in a given stock of fish (Smith et al., 1991). In principle, aquatic reserves (that is, portions of a fishery that are exempt from harvest) can serve as protection against the directional changes in genetic variability that can result from traditional fishing practices, though data to test this prediction are lacking (Roberts and Polunin, 1991).

Shaffer (1987) suggested that three elements will be required for systematic conservation of biodiversity through reserves (1) a classification of the elements of biodiversity, (2) an assessment of the number, size, and location of the reserves, and (3) the active management of the reserve. Each of these three elements is discussed below in terms of genetic theory and aquatic reserve design.

Classification of the Elements of Biodiversity

This first element is perhaps the most difficult one to achieve, yet the one of prime importance. The elements of biodiversity are complex and involve molecular diversity through ecosystem diversity. Surveys of the biodiversity of aquatic organisms at the species level and below need to be conducted in nearly all watersheds to document the genetic resources that contribute to the fishery or the functioning of the aquatic community. In North America much of this information exists or is in the process of being collected for many, but not all, aquatic species. However, in tropical areas and developing countries, basic research and surveys of aquatic species is required (WRI, IUCN, and UNEP 1992).

Historically the elements of biodiversity at the species level and below have been described in terms of morphological variation. The advent of genetic descriptions through analysis of isozymes, chromosomes, and DNA has provided researchers with a powerful tool to assess taxonomic relationships, migration patterns, breeding structure, and the amount and organization of genetic diversity present in aquatic organisms (Ryman and Utter, 1987; Whitmore, 1990). All of these factors will be important in reserve design and management. For example, genetic data on trout populations in Nevada helped establish land use priorities, facilitated land exchange programs, and documented the taxonomic relationship and distribution of native trouts. In addition, these data provided justification for aquatic reserves (Bartley and Gall, 1989).

Because of the vast range in value and popularity of aquatic organisms (compare, for example, lobsters and minnows), aquatic reserves may need to focus on individual species of extreme commercial, social, or ecological importance. These are the "design" or "keystone" species of Moyle and Sato (1991) and the reserve can be designed around the biological requirements of these more "important" organisms. Although we do not wish to advocate a species-specific approach to reserve design, often these favored species will be able to attract popular support and thereby help in the conservation of entire habitats or watersheds (McNeely et al., 1990).

Detailed genetic and population studies of the design species should be a priority for establishing reserves for their conservation. Meffe (1990) described six applications of genetic data to the conservation of endangered species that will also be important in assessing the genetic resources of commercially important species that may be enhanced by aquatic reserves.

- Description of the quantity and geographic distribution of the genetic variation in species
- Estimation of historical levels of natural isolation and gene flow among populations

- Identification of unique gene pools for special protection
- Contribution to taxonomic clarification and stock identification
- Information for choosing stocks to be released (that is, utilized in reserves) through hatchery enhancement or transfers
- Evaluation and monitoring of hatchery populations

Some of Meffe's applications involve assessment and management (see below); fulfilling these criteria may be difficult in developing countries where resources are heavily exploited, technical expertise scarce, and local infrastructure inadequate. Supplies from the international community may be needed in these situations.

Assessment of Number, Size, and Location of Aquatic Reserves

Once the elements of biodiversity are known, to the extent possible with current technology and funding, the actual design and locations of reserves can be initiated to conserve those elements. Reserves can vary from near pristine wilderness to jars of alcohol (Moyle and Sato, 1991). Nyman (1991) and others recommend that aquatic organisms should be conserved where they live, but reserves contain a small fraction of the total area occupied by aquatic communities so they represent some degree of habitat fragmentation and reduction in population numbers. This reduction can be offset by active management (see below) based on ecological and genetic principles.

In small populations genetic variability and vitality can be eroded by inbreeding (the mating of closely related individuals) and genetic drift (the loss of genes due to random processes that occur in small populations). The concept of minimum viable population (MVP) arose to address the question of how large a population needs to be in order to persist for a given period of time (Soulé, 1987). The most widely cited numbers for an effective population size (N_e) that avoids genetic problems are 50 for a managed hatchery population and 500 for a natural population (FAO/UNEP, 1981). N_e represents the size of a population that would lose genetic variation at the same rate as an ideal population with random mating, equal numbers of male and females, and a Poisson distribution in the number of offspring produced (Meffe, 1986). However, these numbers, 50 and 500, are probably good only for an approximate order of magnitude; several workers have demonstrated that N_e may be very much lower than the actual census number N (Waples and Teel, 1990; Bartley et al., 1992).

The size and numbers of aquatic reserves will depend on the goals and design species involved. A debate between whether to have a single large reserve or several small reserves (SLOSS) has subsided (Soulé and Simberloff,

1986), and now the choice of number and size of reserves is influenced by, among other things, the organization of the genetic resources of the design species. If the majority of variability is among populations, as in inland versus coastal species and subspecies of western trout (*Oncorhynchus mykiss* and *O. clarki* ssp.), then multiple reserves throughout the species' range would be needed (Loudenslager and Gall, 1980) to preserve the variety of the species' genetic resource. Conversely, if intrapopulation variability is higher than interpopulation variation, as in the case of some marine mammals (Hoelzel, 1991), then fewer large reserves preserving these intrapopulation dynamics would be required.

A major advance in fishery management has been the incorporation of the stock concept into harvest policy (Gulland, 1969). It is apparent that the aquatic biomass cannot be harvested as if it were all one similar species. The same is true at the species level, because the tremendous amount of genetic and phenotypic variation in stocks or subpopulations within species contributes to the persistence of that species in nature. Aquatic reserves should be established to enhance or complement the natural subpopulation (stock) structure and distribution of genetic variation of a fishery.

For example, the Pacific salmon fishery along the west coast of North America is composed of salmon stocks from hundreds of inland rivers and tributaries. An international genetic analysis has shown that a large component of the total genetic variation is due to genetic differences among drainages (Utter *et al.*, 1989; Bartley *et al.*, 1992). Therefore, reserves to promote the chinook salmon fishery should be established within drainage systems throughout their range. However, the pink salmon (*O. gorbuscha*) have a precise two-year spawning cycle that results in a given river having an odd- and even-year run of salmon that are differentiated genetically (Aspinwal, 1974). Odd year-class runs of pink salmon from different rivers are more similar to each other than are even year-class runs from the same river. Therefore, to preserve the variation in this species, care should be taken to conserve odd and even year classes and to ensure that eggs between the two year classes are not mixed.

Another example concerns several lakes of the Muara Muntai District of Kalimantan, Indonesia, which supply spawning habitat for several species of commercially important riverine fishes. These lakes are closed to fishermen, thereby theoretically preserving the stock structure of the fishery by allowing for natural reproduction, maturation, and migration of the fishes. Mature fish are then harvested in the Mahakam River, which connects the series of lakes (T. Petr, pers. comm. 1991).

Moyle and Sato (1991) stressed the importance of the preservation of outside influences or external connections, including both biotic and abiotic factors, for successful aquatic preserve functioning. Moyle and Sato conclude that intermediate levels of these influences are optimum in order for aquatic communities or reserves to maintain its structure. We can extend this idea to genetic principles by considering gene flow between reserves as an outside influence. Again, depending on the design species, aquatic reserves should incorporate varying levels of gene flow (Meffe, 1990; Ryman, 1992). Many geographically separate populations maintain some small amounts of gene flow through larval dispersal or straying. This seems to be a natural phenomenon that helps populations recover after ecological catastrophes, colonize new areas, and avoid inbreeding. However, populations also are adapted to their local environment, and the introduction of novel or exotic genes would run the risk of breaking down the co-adapted gene complexes or diluting adapted genes. The hope is to define the intermediate level of gene flow such that a population in a reserve maintains adequate genetic variability, but that it does not become swamped by the exotic genes.

Outbreeding depression is the term describing the decrease in population viability that may occur due to the infusion of genes from an external source. The Northwest Power Planning Commission in Oregon (personal observation, 1990 DMB) recommended that controlled experiments be initiated to assess the level of outbreeding depression (if any) that may result from mixing subpopulations of commercially important species of Pacific salmon. This recommendation should be extended to any fishery or area where introductions or the mixing of formerly discrete populations are being considered.

Active Management

All aquatic reserves will require some form of active management, whether it be captive breeding of specific races or ensuring that harvesting regulations are obeyed. Genetic management will include facilitating or preventing the flow of genes between reserves or between reserves and other aquatic areas, the monitoring of the genetic variability of design species, and the management of hatchery populations that may supplement or interact with populations in the reserve.

One form of active management of reserves is the establishment of hatcheries to augment natural breeding. Hatcheries may have either positive or negative effects on natural populations or fisheries through escape or hybridization (Hindar et al., 1991). Knowing the genetic structure of both the fishery stock and the hatchery stock, as recommended above, will facilitate assessment of these effects. Hatchery management practices should incorporate genetic principles to avoid inbreeding through selective matings, to maximize effective population size by using equal numbers of males and females in single-pair matings, and to maintain genetic integrity of races or stocks by avoiding transfers of eggs or mixing of strains (Tave, 1986; Meffe, 1986; Bartley and Kent, 1990). In addition, if hatcheries are meant to supply natural reserves or reserves that are only moderately changed by anthropocentric activities, then care should be taken to avoid domestication and inadvertent artificial selection during the rearing process. This may mean, for example, reducing the time an organism spends in the hatchery.

A serious shortcoming of most aquatic reserves and indeed of many aquatic stock-enhancement programs is the lack of monitoring and accurate assessment of the conservation/enhancement effort. Once reserves are established it will be imperative to assess and monitor the changes in the fishery the reserve was meant to support. This monitoring should involve the same descriptions used to document the elements of biodiversity mentioned above. Although aquatic reserves have been successful in increasing the size and numbers of certain fish resident in the reserve (Roberts and Polunin, 1991), their influence on surrounding fisheries is unclear. For example, in the example of the Mahakam River given above, it is not known what contribution each protected lake makes to the riverine fishery because no stock descriptors or monitoring programs have been implemented. If we err in the design of reserves or their management, accurate monitoring data will at least enable our next attempt to profit from earlier efforts.

For aquatic reserves to function as an effective conservation and enhancement tool, it will be necessary to synthesize information from a variety of sources. Genetic data integrated with other biological and environmental data should provide managers with information to promote the long-term functioning of aquatic reserves.

Side Benefits of Aquatic Protected Areas

Side benefits can accrue when areas are protected for fisheries. These include natural habitat restoration through elimination of illegal and indiscriminate fishing, as by dynamite, and prohibition of fishing methods that may be destructive to the benthos, such as bottom trawling.

Development of Fisheries in Areas Protected for Other Purposes

Fisheries can be reestablished in an area formerly closed to fishing that is being protected for purposes other than fisheries. There are a number of advantages to encouraging fishing in, for example, protected areas such as national parks and wildlife reserves, if the development is well conceived and well managed. Exploitation of fishery resources increases the overall value of the protected area, providing a basis for quantifying the additional income, employment, and contribution to food security available due to fishing. In addition, by "selling" the resources to fishermen, the fishery may be a source of income to help to defray the costs of the protected area. Also, there is an advantage from the viewpoint of fishery administration: Because the fishery exists in an area of controlled access, the fishery is much more manageable and less costly to manage than in a free access situation. Two case studies that well illustrate this situation are the Lac Ihema and Lake Edward fisheries, both within national parks.

Lac Ihema Fishery

Lac Ihema, of 9000 ha, is located entirely within the Akagera National Park in Rwanda. The park was created in 1934. The lake was exploited by about 50 Tanzanian fishermen up until 1980 at which time the exploitation largely ceased. A new development policy to utilize the natural resources of national parks permitted the creation of a fishery on the lake as a project co-financed by the governments of Rwanda and Belgium. A fishery center was constructed on the lake that included lodgings, offices, an ice machine, an insulated truck, and facilities for fish processing.

The fishery is tightly controlled in many ways. Fishermen have to be unmarried. They live at the fishery center for three weeks at a time, then in rotation have one week off. All of the catch is purchased and processed by the fishery center, and the fishing gear and effort are established by the project through its own research program. The ecological integrity of the park is maintained through a number of rules including the prohibition of domesticated animals and of the growing of vegetables. Firewood can be collected, but only dead wood can be used. Areas established as rookeries are closed to fishing.

Economically, the fishery has been a success. Fishermen have a monthly income of up to 10 times the salary of an ordinary worker. About 40 fishermen are employed and supported by 20 additional workers at the center. In 1981 there was a net profit of US$ 15,000 on a gross of US$ 135,000 including a depreciation of US$ 40,000 based on a catch of 168 tons.

The factors contributing to the success of this arrangement are that the initiative to utilize the fishery resources of the park originated with the national parks administration itself and that the same organization is a partner in the fishery. Another factor is the extraordinarily tight control on all aspects of fishing, processing and marketing that ensures minimal impacts on the terrestrial flora and fauna. Finally, the venture proved to be economic as well as providing additional food security for the country. This fishery is still producing 200 tons of fish per year (Frank and Bambujijumugisha, 1982; Mughanda et al., in press; Frank, pers. comm. 1991).

Lake Edward Fishery

The Parc National Albert (now the Parc National des Virunga) was established in 1925. It encompasses the shoreline of Lake Edward (2240 sq km) in Zaire. The lakeside population was evacuated due to the outbreak of trypanosomiasis and was not permitted to return thereafter in order to protect the park environment. Fishing rights were revoked; however, as a consequence of World War II, restoration of fishing occurred in 1943. In 1949 fishing rights were invested in a cooperative for 30 years with the provision that the profits would go to providing common services such as healthcare. A second cooperative was formed to succeed the first, but after a short time most of its operations collapsed and thereafter fishing by independent operators was tolerated by the park authorities.

The Lake Edward situation contrasts in several ways with that of Lac Ihema. Partly it is a matter of scale, 20 units fishing in Lac Ihema compared with 700 on Lake Edward. Also, in Lake Edward, the fishery operates within the park but there are three fishing villages in which agriculture is prohibited and opportunities for supply of goods from outside the park are not good. Twelve guard posts along the lake shore are *de facto* ports, fish processing areas, and markets for the villages just outside the park area that otherwise would be isolated due to poor road access. Although these activities are against regulations, they are permitted for lack of a viable alternative. Another difference is that fishery development has gone from a semi-industrial scale largely dependent on expatriate advice and management skills to an artisanal level with little loss of resource potential. This change has been accompanied by many changes in the organization of fishermen. There are a number of fishery management and development problems, such as fishing on nursery areas and access to population centers outside the park, that are the responsibility of the park administration, but that can be settled in consultation with the various fishermen's organizations. Despite this, it is clear that even in situations where fisheries are large-scale and economically important activities within a park area, the fisheries and national parks can accommodate one another to their mutual benefit (A. Haling, pers. comm., 1990; Vakily, 1989).

Fisheries Affected by Areas Protected for Other Purposes

General guidelines for establishing marine protected areas have been formulated by Kelleher and Kenchington (1991). Problems of implementation of protected areas in which fisheries are affected have already been well reported (Kriwoken and Haward, 1991). Therefore, this section focuses on one of the most important protected area issues with respect to fisheries, that of critical habitats.

Critical Habitats, Fisheries, and Protected Areas

In what framework should critical habitats for fisheries be considered? Odum (1984), looking at coastal habitats and marine fisheries, pointed out that there often is not one critical habitat but a set of functionally connected habitats to be protected. Often, these are along a salinity gradient. Specific coastal habitats critical for reproduction or feeding include areas of upwelling, estuaries, and lagoons as well as seaweed beds, seagrasses, wetlands, and mangrove forests (Caddy, 1990b).

The river basin offers an additional ecological and administrative framework in which to consider critical habitats. Using the basin as a frame of reference helps to define critical habitat interrelationships and to identify areas that might simultaneously be protected for fisheries and other purposes. Advantages are that the concept of river and lake basins for environmental and fisheries management is a familiar one among fisheries workers. It has been promoted internationally for some time (for example, Kapetsky, 1981; Petr, 1982, 1985; Baluyut, 1986). It has already been mentioned that much of the genetic variation among some fishes is accounted for by variation among drainages. Another advantage is that the laterally and longitudinally migrating species of river–floodplain–lake systems that constitute the most important resources of inland fisheries in Latin America, Africa, and Asia are incorporated. From a broader perspective, the river basin accommodates the salinity gradient notion by taking into account anadromous and catadromous fishes.

Critical Habitats or Critical Processes?

Ultimately, critical habitats can be characterized by the natural processes that maintain them. In river systems, the seasonal flooding and recession are the basis of the formation of the critical habitats—the flooded fringe of the plain and the ephemeral streams leading into it are important during inundation. During the recession, the floodplain lakes and river channels provide the critical refuges for survival through the dry season. These are somewhat analogous to coastal seasonally upwelling areas and to offshore areas with seasonal currents.

An important aspect of critical habitats is that they may be affected by quite distant events. This is evident from changes in coastal and estuarine habitats and their fisheries wrought by damming of main-stem rivers. Also, changes to inland seas from water diversion result in increases in salinity and eventual disappearance of fish. More subtle, but nonetheless far-reaching, are the changes in water quality brought about by soil degradation. For example, degraded soils occur over about 17 percent of Africa (Deichmann and Eklundh, 1991). In this way the river basin provides a unifying link between natural and anthropogenic environmental changes inland and their consequences that are felt in coastal and marine waters.

A final word in favor of a river basin approach to critical habitats and protected areas is that, from a political viewpoint, administrative boundaries often follow the natural boundaries of basins. Additionally, from a perspective broader than fisheries and wildlife, there is broad consensus that water resources are best managed at the river and lake basin level. For example, the International Conference on Water and the Environment in its Dublin Statement recognizes the river basin as the most appropriate geographical entity for planning and management at subnational, national, and transboundary basins. This, of course, offers a broad point of contact and communication for protected area agencies.

Identification and Precedence of Protected Areas

For most commercially important species, with the exception of reef fishes (Plan Development Team, 1990), critical habitats can be identified and mapped by using a combination of the available life history and fishery catch and effort data, at the least in a general way and at a large scale (for example, Garcia, 1982; NOAA, 1986); however, this information has been largely under-utilized until recently. Geographical information systems (GIS) can be used as a tool to analyze the information already accumulated. Ideally, aquatic resources information would be synthesized in a way that is akin to that of land resources information systems (Garcia and Kapetsky, 1991). Such a system would have the capability to identify habitats critical to all commercial species and to indicate the degree of coincidence among them. In this way those habitats critical for the greatest number of species, or for the most valuable "design" species, could be pinpointed. Aquatic reserves could then be designed to incorporate them.

The same approach can be used to identify and quantify areas simultaneously critical for other plant and animal life, to identify potentially competing uses such as aquaculture (Kapetsky, McGregor, and Nanne, 1987; Kapetsky and Ataman, 1991), to assess threats to the environment (for example, Trudel et al., 1987), and, in general, to promote multiple-use decision making. An example of an international-scale effort that appears to have the capacity to undertake this kind of analysis comprehensively is the East Coast of North America Strategic Assessment Project (NOAA, 1991). The purpose is to provide scientific information needed to evaluate national and regional development and conservation of coastal and oceanic resources.

The assessment is to have an ecosystem framework. It will deal with estuarine–oceanic interrelationships in terms of species distribution and life histo-
ries. A special effort will be made to understand and quantify habitat loss in
terms of biotic and abiotic requirements of species and the same characteristics of the habitats.

Conclusion

Fishery resources increasingly require protection in order to ensure sustainable exploitation. Because spatial protection of fishery resources has long
been practiced as a management tool, sharing of areas for the protection of
other living resources should be widely acceptable to fisheries agencies. A
strong and compelling reason for creating aquatic reserves is that they will
benefit fisheries. The promotion of the protected area should be directed at
demonstrating the benefits to fishermen.

Aquatic reserves can be based on certain commercially or ecologically
important species. The protection of these species will benefit the entire community or ecosystem.

Incorporating genetic principles into the design and maintenance of
aquatic reserves will promote long-term gains and stability of the fishery
resource. Utilization of fishery resources within areas protected for other purposes can enhance the value of a national park or wildlife reserve as well as
provide income for the operation of the protected area.

The habitat is an important unit for fisheries management because many
habitats are critical to fishery resources. Physical processes such as seasonal
inundation and periodic or continuous openings to the sea ultimately make
the habitats pivotal, so underlying the protection of habitats is the need to
maintain essential physical and ecological processes.

A key point to recognize in identifying protected areas is the control over
not only the resource but also the space occupied by the resource, even if it
is used only temporarily. Habitat use by fishes can occur with diel to seasonal
frequency depending on the scale that defines the habitat. Therefore, the definition of reserve boundaries may have to be dynamic.

For the identification of critical habitats and areas to be protected for
inland fishes and for other living resources, the river basin provides a useful
ecological framework for analysis as well as a systems link with coastal and
marine areas. The river basin also provides an administrative structure for
subnational, national, and international management.

The available information—genetic, ecological, and economic—needs to
be synthesized in order to improve identification and ranking of critical habitats and to design protected areas. Systematic conservation and enhance-

ment by aquatic reserves requires defining the elements of biodiversity; assessment of size, number, and location of protected areas; and active management.

GIS is a technology that lends itself to the diverse kinds of spatial analyses required to identify critical habitats and to design and manage protected areas important for fisheries. GIS is already being implemented cooperatively among agencies to link coastal and offshore systems. To be truly comprehensive and effective, such systems need to include inland waters.

Acknowledgments

A number of FAO colleagues made critical comments on the manuscript. We thank J.F. Caddy, G. Child, T. Do-Chi, P. Martosubroto, T. Petr, A. Tacon, and R. Welcomme. We are grateful to A. Haling and to V. Frank for supplementary information on the Lake Edward and the Lac Ihema fisheries, respectively.

References

Aspinwall, N. 1974. Genetic analysis of North American populations of the pink salmon, *Oncorhynchus gorbuscha*, possible evidence for the neutral mutation–random drift hypothesis. *Evolution* 28:295–305.

Baluyut, E.A. 1986. Planning for inland fisheries under the constraints from other uses of land and water resources: General considerations and the Philippines. *FAO Fish.Circ.* (798): 44 pp.

Bartley, D.M., and G.A.E. Gall. 1989. Biochemical genetic analysis of native trout populations in Nevada. Report on populations collected 1976–1988. Nevada Department of Wildlife, Reno, Nevada. 27 pp.

Bartley, D.M., and D.B. Kent, 1990. Genetic structure of white seabass populations for the southern California Bight region: Applications to hatchery enhancement. *Cal-COFI Rep.* 31:97–105.

Bartley, D.M., M. Bagley, G. Gall, and B. Bentley. 1992. Use of linkage disequilibrium data to estimate effective size of hatchery and natural fish populations. *Cons. Biol.* 6:365–375.

Bartley, D., B. Bentley, J. Brodziak, R. Gomulkiewicz, M. Mangel, and G.A.E. Gall,. 1992. Geographic variation in population genetic structure of Chinook salmon from California and Oregon. *Fish. Bull.*, U.S. 90, 77–100 (authorship amended per errata, *Fish. Bull.* 90(3), iii.

Biro, E. 1992. Fishery reserves—new ways of protecting fish stocks. *National Fisherman* 72 (11):25–26.

Caddy, J.F. 1990a. Options for the regulation of Mediterranean demersal fisheries. *Natural Resources Modeling* 4(4):427–475.

————. 1990b. The protection of sensitive sea areas: a perspective on the conservation of critical marine habitats of importance to marine fisheries. Proceedings on the International Seminar on Protection of Sensitive Sea Areas, Preliminary Edition, Malmo, Sweden 25–28 September 1990. International Maritime Organization, London. 17–31.

Deichmann, U., and L. Eklundh. 1991. Global digital datasets for land degradation studies: a GIS approach. Global Resource Information Database, *GRID Case Study 4*. UNEP, Nairobi. 103 pp.

FAO/UNEP, 1981. Conservation of the genetic resources of fish: problems and recommendations. Report of the expert Consultation on the genetic resources of fish. *FAO Fish. Tech. Paper 217*. Rome, Italy. 43 pp.

Frank, V., and R. Bambujijumugisha. 1982. La pecherie du lac Ihema. *Bulletin Agricole du Rwanda* (4):250–252.

Garcia, S.M. 1982. Distribution, migration and spawning of the main fish resources in the northern CECAF area. CECAF/ECAF Series, FAO, Rome.82/25: 9 pp. + 10 maps.

Garcia, S.M., and J.M. Kapetsky. 1991. GIS applications for fisheries and aquaculture in FAO. Paper presented at "Marine resources atlases—an update." *International Maritime Organization*, London, 17–18 October, 1991. 14 pp.

Gulland, J.A. 1969. *Manual of methods for fish stock assessment*. Part 1. FAO FRS/M4.

Hindar, K., N. Ryman, and F. Utter. 1991. Genetic effects of cultured fish on natural fish populations. *Can. J. Fish. Aquat. Sci.* 48:945–957.

Hoelzel, A.R. (ed.). 1991. Genetic ecology of whales and dolphins. *Report of the International Whaling Commission Special Issue 13*. Cambridge, United Kingdom. 311 pp.

Kapetsky, J. 1981. Seminar on river basin management and development, Blantyre, Malawi, 8–10 December, 1980. Papers presented. *CIFA Tech Pap*. Kay. (8):302 pp.

Kapetsky, J.M., and E. Ataman. 1991. An information base for the orderly development of mariculture and regional project for training on mariculture development and management. Regional Seafarming Development and Demonstration Project, RAS/90/002. *Working Paper SF/WP/91/6*. 45 pp.

Kapetsky, J.M., L. McGregor, and H. Nanne E. 1987. A geographical information system and satellite remote sensing to plan for aquaculture development: A FAO UNEP/GRID cooperative study in Costa Rica. *FAO Tech. Pap*. (287), 51 pp.

Kelleher, G., and R. Kenchington. 1991. *Guidelines for establishing marine protected areas*. IVth World Congress on National Parks and Protected Areas, Caracas, Venezuela 10–21 February, 1992. IUCN, Gland, Switzerland. 90 pp.

Kriwoken, L.K., and M. Haward. 1991. Marine and estuarine protected areas in Tasmania, Australia: the complexities of policy development. *Ocean and Shoreline Management* 15:143–163.

Loudenslager, E.J., and G.A.E. Gall. 1980. Geographic patterns of protein variation and subspeciation in cutthroat trout, *Salmo clarki. Syst. Zool.* 29:27–42.

McNeely, J.A., K.R. Miller, W.V. Reid, R.A. Mittermeier, and T.B. Werner. 1990. *Con-*

serving the world's biodiversity. IUCN, Gland Switzerland, WRI, CI, WWF.US, and the World Bank, Washington, D.C. 193 pp.

Meffe, G.K. 1986. Conservation genetics and the management of endangered fishes. *Fisheries* 11:14–23.

———. 1990. Genetic approaches to conservation of rare fishes: examples from North American desert species. *J. Fish. Biol.* 37(Supp. A):105–112.

Moyle, P.B., and G.M. Sato. 1991. On the design of preserves to protect native fishes. In W.L. Minckley and J.E. Deacon (eds.). *Battle against extinction: Native fish management in the American West*. University of Arizona Press. 155–169.

Mughanda, M., J.C. Micha, J. Degand, and V. Frank, in press. La pecherie Ihema (Rwanda): Production commerciale et rentabilitie socio-economique. Agri-Overseas, AGCD, Bruxelles.

NOAA. 1986. *Gulf of Mexico coastal and ocean zones strategic assessment: Data atlas*. Washington, D.C.: Strategic Assessment Branch and Southeast Fisheries Center. 163 maps and text.

———. 1991. *East coast of North America strategic assessment project: biogeographic characterization component*. Prospectus October 1991. Washington, D.C.: Strategic Environmental Assessments Division. 17 pp.

Noss, R.F. 1991. Sustainability and wilderness. *Cons. Biol.* 5:120–122.

Nyman, L. 1991. *Conservation of freshwater fish*. Fisheries Development Series 56. Swedish Center for Coastal Development and Management of Aquatic Resources. Goteborg, Sweden. 38 pp.

Odum, W.E. 1984. The relationship between protected coastal areas and marine fisheries genetic resources. In *National parks, conservation and development*, J.A. McNeely and K.R. Miller (eds.). Washington, D.C.: Smithsonian Institution Press. 648–655.

Petr, T. (ed.). 1982. Summary report and selected papers presented at the IPFC workshop on inland fisheries for planners. Manila, The Philippines 2–6 August 1982. *FAO Fish. Rep.* (288): 191 pp.

———. 1985. Inland fisheries in multiple-purpose river basin planning and development in tropical Asian countries: three case studies. *FAO Fish. Tech. Pap.*, (265): 166 pp.

Plan Development Team. 1990. *The potential of marine fishery reserves for reef fishery management in the U.S. Southern Atlantic*. NOAA Technical Memorandum NMFS—SEFC-261, 40 pp.

Reid, W.V., and K.R. Miller. 1989. *Keeping options alive, the scientific basis for conserving biodiversity*. World Resources Institute, Washington, D.C.

Roberts, C.M., and N.V.C. Polunin. 1991. Are marine reserves effective in management of reef fisheries? *Reviews in Fish Biology and Fisheries* 1:65–91.

Ryman, N. 1992. Conservation genetics considerations in fishery management. *J. Fish. Biol.* 39 (Supp. A): 211–224.

Ryman, N., and F. Utter (eds.). 1987. *Population genetics and fishery management*. Univ. Washington Press, Seattle. 420 pp.

Salwasser, H. 1991. New perspectives for sustaining diversity in the U.S. National Forests ecosystems. *Cons. Biol.* 5:567–569.

Shaffer, M. 1987. Minimum viable populations: coping with uncertainty. In *Viable populations for conservation*, M. Soulé (ed.). Cambridge, United Kingdom: Cambridge University Press. 69–86.

Smith, P.J., R.I.C.C. Francis, and M. McVeagh. 1991. Loss of genetic diversity due to fishing pressure. *Fisheries Research* 10:309–316.

Soulé, M.E. (ed.). 1987. *Viable populations for conservation*. Cambridge, United Kingdom: Cambridge University Press. 189 pp.

Soulé, M.E., and D. Simberloff. 1986. What do genetics and ecology tell us about the design of nature reserves? *Biol. Cons.* 35:19–40.

Tave, D. 1986. *Genetics for fish hatchery managers*. AVI Publishing Inc. Westport, CN. 299 pp.

Trudel, B.K., R.C. Belore, B.J. Jessiman, and S.L. Ross. 1987. Development of a dispersant-use decision-making system for oil spills in the U.S. Gulf of Mexico. *Application Research Paper 12*. Ross Environmental Research Limited, Ottawa, Ontario, Canada. 22 pp.

Utter, F., G. Milner, G. Stahl, and D. Teel. 1989. Genetic Population Structure of Chinook Salmon in the Pacific Northwest. *Fish Bull.* U.S. 87:238–264.

Vakily, J.M. 1989. Les peches dan la partie zairoise du Lac Idi Amin. Departement de Affaires Foncieres, Environnement et Conservation de la Nature. Rapport Technique des Peches au Zaire. Kinshasa. 48 pp.

Waples, R.S., and D.J. Teel. 1990. Conservation genetics of Pacific salmon 1. Temporal changes in allele frequency. *Cons. Biol.* 4:144–156.

Whitmore, D.H. 1990. *Electrophoretic and isoelectric focusing techniques in fisheries management*. Boston: CRC Press, 350 pp.

WRI, IUCN, and UNEP. 1992. *Global Biodiversity Strategy.* World Resources Institute, Washington, D.C. 244 pp.

Chapter 13

Protected Areas and the Hydrological Cycle

Patrick Dugan and Edward Maltby

Editor's Introduction:
Water is one of the essentials for human existence, and protected areas have a major role to play in ensuring that water resource development is a positive part of social and economic development. As described by Patrick Dugan, IUCN's Director of Regional Affairs Division, and Edward Maltby, from the University of Exeter, UK, protected area managers can do much to build positive relationships with agencies interested in water resource management. Abundant evidence is now available of the value of protected areas in managing catchments and in ensuring a regular supply of high-quality water, but much remains to be done to promote appropriate investment by governments in protected area management as an integral component of water resource development. Demonstrating this role for each protected area will help build support from both national and international agencies and establish productive working partnerships with institutions and individuals having responsibility for water resource management, as well as other resource users downstream. The water resource management agencies themselves should also recognize that protected areas have much to contribute to their own interests.

Introduction

For much of the past two decades the constant threat of drought has dominated natural resource management policy in many parts of the world. In some, notably in the Sahel, the development of coherent national and regional responses to the vagaries of annual rainfall has been the central priority of governments and donors alike.

The industrial world has not been immune to this crisis, with 1991 bringing the fifth year of severe drought in California; agriculture in the Mediterranean basin has been constrained by a sequence of dry summers, falling

groundwater, and reduced riverflow; and even the United Kingdom is growing concerned over the consequences of progressively lowering groundwater
levels.

Ironically, images of drought-stricken farmers on our television screens are
frequently followed by floods a year or even a few months later. In August
1988, only a few years after the tragic scenes of drought and starvation in the
Sahel, the world's attention was again drawn to the region, but this time by
floods. In Nigeria dams burst, and in Sudan the swollen Nile flooded large
areas of Khartoum.

In an effort to alleviate the human suffering and economic crises caused by
drought and flood, governments and the development assistance community
have in recent years invested billions of dollars in building dams, dikes, and
otherwise regulating rivers. In northern Nigeria 16 dams have been built on
the Hadejia Jama'are river system alone (van Ketel *et al.*, 1987), with the goal
of harnessing the river's water resources for agriculture and urban use.
Nearby a steady decline in rainfall has reduced the flow of the Logone and
Chari rivers and resulted in the gradual shrinkage of Lake Chad. In response,
a proposal to transfer water from the Zaire basin to Lake Chad has been studied by the Lake Chad Basin Committee and promoted by foreign engineering
companies. In Asia the preliminary studies of the Flood Action Plan designed
to limit Bangladesh's devastating floods are being pursued under the coordination of the World Bank.

Ironically, as this major international investment is being made in structural approaches to water management, limited attention has been given to
the role of natural ecosystems in managing the hydrological cycle and to the
potential for improved management of natural aquatic ecosystems and catchment basins as alternatives to major engineering investments. Yet as early as
1972 the value of nonstructural approaches to managing water flow was recognized by the US Army Corps of Engineers, who recommended that the
most cost-effective approach to flood control in the Charles River of Massa-
chusetts lay in preserving the 3800 ha of mainstream wetlands that provide
natural valley storage of flood waters. Yet only rarely does this recognition of
the economic value of the hydraulic function of natural ecosystems feed
through into land use policy and development investment. While the past 20
years have seen a laudable increase in the funding that has gone to catchment
management as a component of dam construction, efforts to maintain the
hydrological processes on which the dams depend are rarely of the magnitude
required.

Perhaps even more important, in the absence of a full appreciation of the
value of natural ecosystems in managing the hydrological cycle, many of the
current investments—notably dams, dikes, and river alteration—lead to
many forms of ecosystem degradation downstream, exacerbate the effects of

drought, and reduce options for meeting the social and economic development needs of the rural poor.

In the face of these problems there is growing recognition of the need to develop a more broad-based approach to water management. While dams and dikes will continue to be built, these form but one of many options for managing freshwater resources. Natural hydrological systems and the landscapes that sustain them must be managed in a way to maximize the multiple values of these water resources; this will require substantial investment in managing catchments and floodplains and estuaries downstream. In doing so protected areas are one of the most valuable management options available and can serve as catalysts for environmentally sound management over a wider area. To achieve this, however, protected area managers will need to build collaborative links with hydrologists, hydraulic engineers, and water resource planners. The present chapter reviews these issues, examines the form that such protected areas can take, and underlines the need for partnerships with other disciplines.

Managing Catchments

By the end of the present decade, 12 African countries with a total population of approximately 250 million people will suffer severe water stress. With increasing population, 10 other African countries will be similarly stressed by the year 2025. Approximately 1100 million people, or two-thirds of Africa's projected population by that time, will then live in these 22 countries, and four (Kenya, Rwanda, Burundi, and Malawi) will be facing an extreme water crisis (Falkenmark, 1989).

While the severity of the water crisis now facing Africa is of special significance because it is already so widespread across this most arid of continents, a water resource crisis is now widespread in both the industrialized and the developing world. From California to Bangkok, and India to Honduras, the limited availability of clean water is now seen as a major constraint to further social and economic development. In the Middle East many commentators argue that the likely catalyst of any new regional conflict will be the need for fresh water.

In response to this growing crisis there is today widespread recognition of the need to invest substantially in the management of freshwater resources. Summing up this need, *Caring for the Earth* (IUCN/UNEP/WWF, 1991) has called for specific action.

- An improved information base for sustainable water management

- Better awareness of how the water cycle works, the effect of land uses on the water cycle, the importance of wetlands and other key ecosystems, how to use water and aquatic resources sustainably, and better training in these matters

• Management of water demand to ensure efficient and equitable allocation of water among competing uses

• Integrated management of all water and land uses

• Improved institutional capacity to manage fresh waters

• Strengthened capacity of communities to use water resources sustainably'

• Increased international cooperation on water issues

• Conservation of the diversity of aquatic species and genetic stocks

In particular *Caring for the Earth* stresses that long-term management of water resources will require not only that water demand be managed to ensure efficient and equitable allocation of water among competing uses, but also that an effective series of actions are taken to ensure a regular flow of high-quality fresh water. This in turn will require giving greater emphasis to the drainage basin as the unit of water management, effective integrated management of these catchments, and conservation of critical habitats with important hydrological functions. This can be valuable for ecological, hydrological, and economic reasons. Thus Mackinnon (1983) has shown that the cost of establishment of protected areas, reforestation where necessary, and other measures to protect the catchments of 11 irrigation projects in Indonesia ranged from less that 1 to 5 percent of the development costs of the individual irrigation projects. This compares very favorably with the estimated 30–40 percent loss in efficiency of the irrigation systems if catchments were not properly safeguarded. And in Honduras, the La Tigra National Park, 7500 ha of cloud forest, sustains a high-quality, well-regulated water flow throughout the year, yielding more than 40 percent of the water supply of the capital city Tegucigalpa. Because of its value for watershed protection, La Tigra is today the focus of an investment program involving a series of economic incentives for villagers living in the buffer zones (McNeely, 1988).

In some cases engineers are recognizing earlier mistakes and are prepared to recommend the high costs of putting them right (Maltby, 1986). In the 1960s, the US Army Corps of Engineers built a US$ 29 million canal along Florida's Kissimmee River to control seasonal floods that washed over the riverbanks and damaged property and farmland. Flooding had been particularly severe in 1947 and 1948. But the canal had a disastrous effect. Some 8000 ha of marshland was lost, resulting in a major decline in wildlife. Of greater economic importance was the wide impact on water quality. Before the canal, the naturally sinuous course of the Kissimmee slowed floods resulting from storms. The marginal wetlands not only helped control floods, but acted as chemical filters on the water passing through them. After the canal was built, farmers moved quickly onto the new protected floodplain. More water was running off the areas because there was no wetland vegetation to slow it; this runoff carried large amounts of fertilizer from the farmland, and oxygen levels in Lake Okeechobee were quickly depleted.

A 1972 University of Miami report concluded that "the canal was a major factor in accelerated lake eutrophication, with resultant water quality deterioration." It recommended halting the discharge of all waste materials into the basin and developing a plan for reflooding the marshes of the Lower Kissimmee Valley (quoted in Horowitz, 1978). In 1976 the State Legislature passed the Kissimmee River Restoration Act, and the South Florida Water Management District is now preparing to spend more than it originally cost to build the canal to return the river to its original course.

"Our goal," stated Florida Governor Bob Graham, "is that by the year 2000, the water system will look and function more as it did in the year 1900 than it does today" (Angier, 1984). Despite the delays in its implementation, the decision points the way to wider acceptance of the vital role of natural wetlands in the economic and ecologically sound management of water resources.

These are encouraging trends but today remain the exception. While environmental guidelines for dam construction certainly exist, and the bilateral and multilateral agencies are taking increased note of these, such efforts remain for the most part compensatory add-ons rather than being part of a fundamentally new approach to managing water. For example, in Indochina proposals for development investment in the Mekong Basin list a wide range of water management projects, yet there is so far little accompanying investment to protect the catchment. In the catchments of Africa's great rivers only tentative steps have been taken to establish protected areas that might contribute to stabilizing land use and help restore a regular and predictable supply of fresh water.

Thus while there is now abundant evidence of the value of protected areas in managing catchments and in ensuring a regular supply of fresh water, much remains to be done in order that governments and the development assistance community view and invest in protected area management as an integral component of water resource development. To achieve this, protected area professionals will need to engage in a wider dialogue with hydrologists and water engineers to understand more fully their concerns and to identify areas of mutual interest. Similarly, governments need to pursue a more holistic approach to water management, one that recognizes the role of natural ecosystems in controlling hydrological processes and the role that a well-managed hydrological regime can play in supporting productive natural ecosystems downstream.

Increased investment is needed to expand the coverage of catchments by protected areas and to manage them more effectively. In Nigeria, when the Tiga dam was built at a cost of several million dollars, only a tiny fraction of this sum was allocated to management of Bakolori Game Reserve, which protects much of the catchment, and there has been no major attempt to stabilize land use in the surrounding area. There are numerous other examples where these first steps have been taken but never followed through. In many instances this is a reflection of the inadequacy of the protected area system in the country in question. However, by demonstrating the important role that catchment management can play in meeting the water crisis, protected area managers can establish an important case for securing the national and international support required to strengthen their work.

Despite their importance, protected areas need to be seen as but part of the solution to catchment management. In most areas they will cover only a relatively small proportion of the entire catchment. However, by maintaining natural vegetation cover and fulfilling a range of protective ecosystem functions, protected areas can serve as focal points for further investment in sound land management and as economic investment in the buffer zones and the wider catchment. In other situations the catchment will already have been so modified by people that establishing protected areas would be an inappropriate approach to management. However, by emphasizing the importance of integrated catchment management, protected area managers can play a major catalytic role in encouraging the development of the fully integrated approach to land and water management required to meet the water crisis.

The Water Crisis and Aquatic Ecosystems

While the highly visible scenes of desiccated water holes, drought-stricken crops and cattle, and polluted urban discharge epitomize the world's water crisis in most people's eyes, the degradation and desiccation of freshwater resources have far-reaching impacts on a wide range of natural ecosystems and human activities dependent on them. For example, in West Africa the persistent drought in Senegal and the Fouta Djallon massif of northern Guinea has reduced flow in the Casamance River to the extent that salinity

100 km upstream is now several times that of seawater. Drought in Cameroon reduced flow in the Logone River for most of the 1980s (Drijver and Marchand, 1985).

These natural phenomena have been exacerbated, in some instances substantially, by the construction of dams and diversion of water, designed to make "maximum productive use" of these resources. For example, in northern Nigeria drought has reduced flow in the Hadejia River by 23 percent, but construction of Tiga Dam has reduced river flow by a further 23 percent (Adams and Hollis, 1988). In neighboring Cameroon construction of Maga Dam in 1979 has further reduced flow of the Logone River and limited inundation of the floodplains of Waza National Park.

In Pakistan the flow of the Indus River has been reduced drastically by construction of dams and irrigation schemes. Today only 20 percent of the river's 150 million acre feet (MAF) reach the delta, while proposals exist to cut this further to about 7 percent. Water diversion in Russia has combined with a period of drought to reduce the Aral Sea to 60 percent of its former area and 33 percent of its volume over the course of the past 30 years (WRI, 1990).

These investments in river diversion were designed to maximize agricultural productivity. Yet many of the benefits have fallen short of expectation while productivity of natural systems, which are of equal and often greater importance to the local and national economy, have been reduced. For example in Nigeria the value of agriculture, fisheries, and forestry on the floodplain of the Hadejia river has been calculated as exceeding that of the government-organized irrigated agricultural operations (Barbier et al., 1991).

In Cameroon the loss of the floodplain has devastated fisheries, severely disrupted the pastoral economy, and contributed to the degradation and reduced tourist potential of Waza National Park. Yet with careful management these natural systems provide multiple economic alternatives to the intensive production of irrigated rice, which is vulnerable to the vagaries of world market price and the limitations of intensive irrigated agriculture in the Soudano–Sahelian zone (Adams, 1992).

These examples do not of course demonstrate that all dams are necessarily negative in impact, but rather illustrate that major investments in an effort to harness water resources are often poor investments. Instead, it is frequently both socially and economically preferable to invest in integrated management of those ecosystems that depend for their productivity on natural freshwater flow.

Such integrated management is a complex process and faces many difficult challenges, notably in establishing systems of resource use and control that are adapted to the unpredictable riverflood and the increasing demand of local populations. In doing so, much can be learned from traditional systems of resource management, which have been pursued by many rural societies for centuries (Adams, 1992). Today, the equilibria established through these traditional systems have been disturbed by increased populations, drought, and changing administrative practices. However, more needs to be invested in trying to establish new equilibria rather than opting immediately for large-scale structural alternatives, which to date have proved notably unsuccessful (Adams, 1992). In many instances these approaches can be combined with dams and other structural approaches, provided that these are designed to maintain at least a minimal river flow. Such approaches are now being explored by several governments, and the government of Senegal has accepted proposals for such a controlled flood as part of its long-term strategy for management of the Senegal River (Horowitz, 1990).

Protected areas have a central role to play in developing approaches to integrated management of these floodplain systems. They can help conserve and manage sites of special importance on the floodplain; by developing access rules that regulate use of resources in time and space, they provide a clear mechanism for regulating use of critical areas such as fish nurseries, floodplain forests, or pasture at those times of year when exploitation would degrade the resource.

Greater importance needs to be given to the maintenance of water flow when management plans are prepared for these protected areas. All too often, however, protected area managers are content to lament the limitations placed on management because of the reduced water flow. Yet recognition of the dependence of protected areas on water flow requires that protected area managers broaden their management horizon and seek to work with other institutions and individuals who have responsibility for water resource management, and with other resource users downstream.

Many coastal systems are equally vulnerable to reduced freshwater inflow, and these impacts merit careful assessment before investing in diversion. Protected areas can provide a central element of integrated management strategies for both coastal zones and floodplains.

Conclusion

Life on Earth depends on water. Yet our use of water has created a crisis for much of the world. As efforts to address this crisis increase there is a growing appreciation that while water demand needs to be controlled much more effectively, major investment is needed to maintain the ecological and hydrological integrity of the drainage basins that yield our freshwater resources. Only a few years ago fresh water was viewed almost exclusively as a commodity that must be harnessed before yielding benefits for human society. There is now growing understanding of the critical role that fresh water plays in maintaining a wide range of economically as well as environmentally important ecosystems, and in promoting increased investment in their management.

As governments and the development assistance community respond by increasing investment in management of the drainage basin, protected area professionals have a central role to play. Protected areas are frequently the most appropriate mechanism for managing the catchment basin in order to ensure a sustained flow of fresh water, or to harness maximum benefit from freshwater ecosystems downstream; thus protected area staff should be well placed to argue forcefully for more integrated management of land use in catchment basins, and for integration of the development of water resources with conservation of the ecosystems that play a key role in the water cycle. However, if protected area staff are to play this role, training institutions will need to do more to provide them with the training and information tools required for them to initiate this dialogue.

This broadening of competence will only be truly effective if the protected area institutions embrace forcefully their role in managing fresh water. This will require greatly enhanced partnerships with departments of water resources and other institutions that have a specific mandate to manage each country's water resources. Failure to do so will limit the contribution of protected areas to addressing the water crisis, and in turn will remove one of the major arguments for further strengthening protected area institutions in the future.

References

Adams, W.M. 1992. *Wasting the Rain*. London: Earthscan.

Adams, W.M., and G.F. Hollis. 1988. *The Hadejfia-Nguru Wetlands Project*. Mimeographed report to IUCN, ICBP, and RSPB. 181 pp.

Angier, N. 1984. "Now you see it, now you don't." *Time*, 6 August 1984.

Barbier, E., W. Adams, and K. Kimmage. 1991. *Economic Valuation of Wetland Benefits: The Hadejia-Jama'are Floodplain, Nigeria*. London Environmental Economics Center, London.

Drijver, C.A., and M. Marchand. 1985. *Taming the Floods: Environmental aspects of floodplain development in Africa*. Center for Environmental Studies. University of Leiden.

Falkenmark, M. 1989. The massive water scarcity now threatening Africa—Why isn't it being addressed? *Ambio* 18(2):112–118.

Horowitz, R.I. 1978. *Our Nation's Wetlands*. Council of Environmental Quality, U.S. Government Printing Office.

IUCN, WWF, and UNEP. 1991. *Caring for the Earth: A Strategy for Sustainable Living*. Gland, Switzerland. 228 pp.

MacKinnon, J.R. 1983. *Irrigation and Watershed Protection in Indonesia*. Report to the World Bank.

Maltby, E. 1986. *Waterlogged Wealth: Why waste the world's wet places?* London: Earthscan. 200 pp.

McNeely, J.A. 1988. *Economics and Biological Diversity: Developing and Using Economic Incentives to Conserve Biological Diversity*. IUCN. Gland, Switzerland. 236 pp.

van Ketel, A., M. Marchand, and W.F. Rodenburg. 1987. *West Africa Review. Edwin Report No. 1*. Center for Environmental Studies, University of Leiden, The Netherlands. 48 pp.

WRI. 1990. *World Resosurces: 1990–91*. WRI. New York. 383 pp.

Chapter 14 ───────────────────────────────

Protected Areas as a Protection Against Natural Hazards

Stephen O. Bender

Editor's Introduction:
Natural events such as earthquakes, volcanic eruptions, tsunamis, landslides, floods, and drought pose hazards for protected areas. On the other hand, the ecosystems in these areas often contain naturally occurring habitats such as reefs, mangroves, dunes, marshes, and forests that reduce the impact of natural hazards on the human population and on economic and social infrastructure. Natural habitats are also affected by development activities, many of which are directly dependent on the structure and function of the particular protected area. In the final analysis, the objectives of protected areas should include the management of these naturally occurring mitigating elements. And the design and management of the protected areas themselves should include components for limiting damage, emergency preparedness, and disaster response when natural hazard events occur. By helping to mitigate natural hazards, protected areas make an important economic and social contribution to human welfare. This role needs to be recognized by those responsible for mitigating such hazards and by society at large. As Stephen Bender, a specialist on environmental management at the Organization of American States, points out, natural hazard management is part of protected area management, and a potentially productive partnership exists between the emergency preparedness and disaster response community and protected area managers.

Introduction

A direct, if not much discussed, relationship exists between disasters, environmental management, and development. In Latin America and the Caribbean, growing population pressures and the demands made on the region's ecosystems have long outstripped their capacity to provide sufficient goods and services in a non-value-added condition, including those areas currently or destined to become protected areas.

Areas safe from natural hazards, pollution, and accidents must often be engineered to be so. But far too often, disaster mitigation is limited to on-site design that at best allows for the passing on of the hazard. In the worst case, the inhabitants or users have neither the technical nor the financial means to reduce the vulnerability of even their own endeavors. On the other hand, protected areas often contain natural hazard mitigation elements that can benefit local people.

To be able to continue providing safe areas, environments must be managed in an integrated way, including the recognition of the impact natural events have on those environments, in terms of both positive attributes and hazards. Natural events and the hazards they pose are part of the systems that make up our environments. On the benefit side, natural events shape the topography, deposit volcanic soils, flush estuaries, water the land, expose buried resources, dispose of combustible material, and continually reset regenerating cycles into motion.

Natural hazards are part of "environmental problems" in every sense of the word. They affect the habitats and species, make manifest the alteration of natural systems, and heighten the impact of those systems' degradation. They spread in an uncontrollable way the results of humans spoiling their environments (Bender, 1992). The impacts of natural disasters together with civil strife, famine, and epidemics are playing an ever-increasing role in setting the development agenda for many developing countries.

This chapter discusses the interrelated aspects of protected areas and natural events that pose hazards.

The Role of Natural Areas in Mitigating the Effects of Natural Hazards on Human Populations, Economic and Social Infrastructure

Protected areas are effective in mitigating the negative impacts of natural events on human populations in a number of ways if the structure and function of the natural systems are maintained and managed effectively. The case for using protected areas as buffer zones for natural hazards is twofold. First, these areas provide naturally occurring mitigation elements that help prevent losses in human populations, economic production capacity, and service infrastructure. Second, these same elements, such as reefs, dunes, and forests, are the very attractions that prompt and deserve conservation. Mangrove trees contribute to the protection of shorelines from the coastal erosion that results from ongoing wave action or coastal erosion that is caused by severe storms (such as hurricanes with associated wave and wind action, storm surge, and flooding).

Mangroves, important parts of estuarine systems, play vital roles in protecting neighboring ecosystems and marine species. Mangroves buffer salinity changes of coastal areas and stabilize riverbed sediments by reducing the sediment load in the water column and then trapping and binding these sediments (Birkeland, 1983). Mangroves reduce soil erosion and maintain the quality of coastal waters by slowing the flow of silt and letting particles settle out (Salm, 1984).

In addition to harboring diverse marine life, mangroves often act as mitigators in natural disasters by lessening their effects on coastal areas. Much like coral reefs, mangroves are natural "breakwaters." During tropical and subtropical storms they dampen waves and high winds, and lessen the impact of tsunamis by absorbing part of the energy of these giant walls of water (Salm, 1984).

Coral reefs are cherished more for their beauty and biological richness than for the subsistence and security they provide to coastal communities around the globe. This approach jeopardizes the very existence of several coral reefs in the Caribbean as well as in Central and South America. Coastal reefs are the centers of primary productivity. They grow in nutrient-poor waters, trapping the few materials that are carried by ocean currents, and introducing energy-rich organic substances into the oceanic food web. They serve as aquatic food producers, luring and nourishing fish which in turn provide essential protein to feed fishermen and their families.

Reefs are also a prime source of marine invertebrates and plants whose potential as a source of food and medicinal products is unlimited and not yet fully explored. Furthermore, coral reefs provide physical protection to low tropical coastlines against damaging waves, tidal surges, coastal erosion, and storm damage. As an example, most of the barrier beach systems studied on St. Croix, U.S. Virgin Islands, was protected by offshore reefs. Yet the reefs' presence shows the strength of the forces that shaped them. Hurricane Hugo pushed these features landward, and future storms may do likewise. Houses and parking lots located near the high tide line of the beaches were most damaged by the storm. Coastal structures and practices to defend them from marine advance tend to disrupt the processes that control the development of barrier systems and ensure the continued presence of natural beaches.

Barrier beaches serve an important protective function for shoreward ecosystems and societal infrastructure by dissipating storm wave energy and by blocking storm-elevated sea levels. Plant communities are important to both the development of barrier landforms and these landforms' effectiveness as storm buffers. The types, composition, and distribution of coastal plant communities are largely controlled by elevation and hydrology, substrata composition, and exposure to wind and salt. Because of their position and role in shoreline protection, these systems often bear the brunt of the storm's force and exhibit significant storm impacts.

Beach, dune, and marsh vegetation reduce damage to the barrier landforms. In addition, natural and anthropogenic toxins are filtered out in these lowland areas. Specifically, vegetation reduces erosion and collected overwash deposits on dunes, and protects landward vegetation. An islandwide survey (Rogers, McLain, and Tobias, 1990) indicated that the areas where coastal development was affected most severely by storms were not protected by naturally vegetated coastal landforms.

Losses resulting from two recent hurricanes in the Caribbean—Gilbert in 1988 and Hugo in 1990—illustrate the importance of protecting natural areas through hazard mitigation. The negative impacts of these two hurricanes on the tourism industry of a number of Caribbean and Central American countries was significant (OAS, 1991; Vermeiren, 1989). Short-term effects included damage to hotels, roads, public works, and the homes of people who work in the service sector of the tourist industry. Poor siting, design, and construction contributed to other losses. For example, in Cancun, wave and wind action from Hurricane Gilbert destroyed hotels that were not anchored to the bedrock. Other hotels suffered severe damage because of poor siting and orientation. Lack of attention to design and detailing during construction contributed to severe losses at other hotels. Although there were few deaths in the Cancun area, the economic disruption in the tourism sector was significant (NAS, 1991b).

Beyond the direct losses in infrastructure and ability to provide services for tourism, losses to the ecosystems in the Yucatan from Hurricane Gilbert were significant. Thousands of flamingo fledglings drowned and other animal and plant species were substantially affected. Due to anthropogenic activities on the barrier islands, including tourism facilities and inadequately designed and constructed infrastructure, the geologic structure of the barrier islands was altered and hydrologic processes were changed. These effects contributed to the breaching of the barrier islands, which in turn resulted in saltwater intrusion and profound effects on plant and animal communities in the area (Clark, 1988). Storm surge and flooding that would otherwise have retreated quickly from the barrier islands was instead caught and pooled for long periods, causing deaths to species and losses to ecosystems. Scientists have observed continued negative impacts during the past three years on plant and animal communities in the area.

In the Ganges-Brahmaputra River Delta of Bangladesh the extent of cyclone damage behind the Sundarbans mangrove forest has historically been less than behind nonmangrove coasts. With this in mind, the Bangladesh afforestation program was developed with hope that the mangrove plantations would provide timber resources, offer coastal agricultural land, and protect nearby villages from storms (Saenger *et al.*, 1983).

A 1990 storm that hit the Sitakunda area vividly illustrated the degree to which mangroves can lessen the effects of cyclones. Several months prior to the storm a 2 km long seawall (10 m high) of steel concrete blocks had been constructed. During the storm about 25 percent of the seawall was smashed and some blocks were moved 100 meters inland. Seaward of the wall part of the shoreline was occupied by 0.95–1.7 m high mangrove trees, which were scarcely damaged by the storm (less than 1 percent of the trees suffered any damage). Coastal areas that did not have concrete walls and were protected by mangroves alone, suffered no more damage than areas with the wall (McConchie, 1990). McConchie posits that the mangroves suffered very little damage because they possessed a degree of flexibility and, unlike the seawall, did not offer a rigid barrier to the wave and current action.

The Effects of Natural Hazards on Ecosystems

Hazard-mitigating natural systems are themselves vulnerable to the destructive forces of nature. The protection and management of protected areas should include the technical and temporal dimensions of hazard impact.

For example, siting hotels slightly inland off barrier islands helps protect the ecosystems of the islands. This in turn maintains the natural resilience of the barrier islands when affected by hazardous natural events. Where the geologic structure and biological composition of the barrier islands are not significantly altered by development, the natural systems are not only less prone to damage or alteration, but also are more effective in mitigating the effects of natural hazards on human populations and on social and economic infrastructure. Such actions also makes them less vulnerable to storm surge, flooding, and other potentially damaging effects of severe storms and hurricanes.

Violent storms, such as hurricanes, are important structuring agents that influence species compositional development, and in some cases the natural disturbance may be beneficial to the ecosystems. For example, seasonal flooding may be an important part of the process of reintroducing nutrients into areas, and hurricanes may improve forest stands. Crow (1980), in studying a tropical rainforest in Puerto Rico, suggests that periodic disturbance is an important factor in the development of these forests. For example, winds act as thinning agents removing many of the standing dead trees. These newly created gaps promote advance regeneration while providing suitable habitat for seedling establishment. Hurricanes may introduce new sources of genetic material (Lugo et al., 1981), which may increase the species richness of a depauperate ecosystem. However, magnitude, intensity, duration, and frequency are important variables of natural disturbance regimes and may prove to produce significant losses to the ecosystems.

The eye of Hurricane Hugo, one of the strongest storms in the Caribbean this century, passed directly over St. Croix, causing widespread devastation. The storm caused widespread destruction in the Virgin Islands National Park and Biosphere Reserve on St. John, the site of numerous research studies on coral reefs, reef fishes, seagrass beds, mangroves, and the moist and dry forests. The storm uprooted and defoliated trees and destroyed large portions of shallow coral reefs. Sustained winds of 225 kilometers per hour and gusts up to 321 kilometers per hour battered the islands for more than 12 hours (Case and Mayfield, 1990). Large trees toppled over, others were stripped of their leaves, and portions of beaches were washed away.

Damage to both marine and terrestrial systems from Hurricane Hugo was "patchy" on several scales. This variability in the amount of damage sustained from major storms had been noted in other studies of reefs (Woodley et al., 1981). It is not clear if St. John's forests or marine systems will recover fully; it is clear that recovery will be delayed by natural processes such as additional storms and coral diseases, as well as activities such as dredging, hillside clearing (which accelerates runoff), destructive anchoring, and boat groundings.

The Vulnerability of Naturally Occurring Mitigation Elements to Anthropogenic Change

Studies are underway to quantify differences in natural and anthropogenic disturbance (natural hazards and anthropogenic activities) to ecosystems in Yucatan (Oaks and Savage, 1991). Anthropogenic disturbance may be more significant than natural disturbance. An example is tourism, which is vital to the economies of many regions possessing attractive coastal and upland areas, particularly the countries of the Caribbean basin. But tourism development threatens the ocean, wetland, and mountain ecosystems it promotes, thus decreasing these elements' mitigating role in lessening natural disasters. The scale of anthropogenic disturbance in global processes is so significant that no ecosystem may be entirely immune to it. Recent scientific assumptions about global environmental change caused by carbon dioxide emissions, the resultant effects of global warming, and the associated sea level rise (along with increased severity and variability of natural hazards such as hurricanes), make global ecological management as important as the management of local ecosystems. While the extent and degree of anthropogenic disturbance must be managed at many scales (such as global, regional, subregional), the most realistic management strategies at the local and regional levels combine preservation and conservation with development investment.

The vulnerability of ecosystems to anthropogenic change is a significant issue for coastal and upland ecosystems, but adequate planning and management of protected areas can help prevent overuse and destruction. In previous decades, generally accepted resource management strategies included the prevention of naturally occurring disturbances (for example, wildfire suppression). Disasters such as the Yellowstone National Park fire in 1988 in the United States illustrate that greater losses were incurred due to fire suppression management strategies because a greater amount of natural fuels had accumulated through time and contributed to the magnitude of the disaster (NAS, 1991a). Other recent lessons about human intervention in the Everglades National Park of Florida point to the shortcomings of management strategies that do not include biological time scales. Drastic changes in the natural hydrologic cycles within the park and hundreds of kilometers outside its boundaries were made through human intervention, contributing to eventual damage for the plants and animals that inhabit the Everglades.

It is well known in Jamaica that careless deforestation of watersheds has led to soil erosion and the loss of thousands of tons of sediment to the sea. Less well known is the fact that this process, facilitated by the destruction of mangroves and other coastal sediment traps, has also degraded Jamaica's coral reefs, stressing or killing corals with sediment or turbidity. This terrestrial runoff carries excessive dissolved nutrients magnified by sewage and fertilizers. Without the coastal wetlands to process these nutrients, they flow out over the reefs, where they facilitate the growth of algae (O'Callaghan, Woodley, and Aiken, 1988).

Although most pollutants are now under governmental regulation, the destruction of reefs through more direct anthropogenic activity is harder to assess and control. Such activities include commercial as well as recreational uses. Construction of tourist facilities, research facilities, and navigational aids has an impact on the reefs. Such development may alter water flow around the reef, thereby changing a major ecological factor. These facilities may also partially shade the reef, reducing photosynthesis.

Natural disturbance of ecosystems is significant even without anthropogenic disturbance. Therefore, it is doubly important to manage natural

areas, such as protected areas, so that the effects of anthropogenic disturbance will not exacerbate the effects of natural disturbance, as happened in
the cases of Hurricanes Gilbert and Hugo.

The Need for Hazard Mitigation as Part
of Protected Areas Planning and Development

In the final analysis, the manifestation of the relationship of protected areas
to natural hazards must be visible in their planning and development. The
role of these areas as naturally occurring mitigating elements must be
included in a description of their purpose, and in their defense for creation,
funding, management, and development. At the same time, it must be recognized that these same protected areas are subject to damage, not only from
human-induced factors, but also from the very events they may, on occasion,
successfully mitigate. For that reason, the design and management of protected areas should include components for mitigating damage, emergency
preparedness, and disaster response.

Mitigating damage from natural hazards begins with an assessment of the
events that could affect an area and the resulting vulnerability. That vulnerability refers not only to the inherent structure and function of the ecosystems contained in the protected area, but also to the infrastructure, both
social and economic, that may be planned and developed as part of the area's
management plan.

As mentioned earlier, the location, severity, and frequency of events that
can affect an area should be understood to the greatest extent possible. In this
way, a determination can be made of anticipated alteration of the area that
goes beyond normal and/or desirable impacts from storms, drought, volcanic
eruptions, and so on. A determination can also be made of the probable
impacts to the facilities, transportation access, signs, and other infrastructure
that may be part of a protected area management scheme. This is particularly
important if guided tourism, scientific research, or general tourist access is
allowed.

As with assessing other areas of social and economic value, the level of
acceptable vulnerability of protected areas is related to the social, financial,
and economic costs and benefits associated with altering the expected events,
or mitigating their impact. The probability of occurrence of a devastating
earthquake or drought affecting a particular protected area may be determined to be so low that it can be assumed, for the life of the proposed investments and management plan, that naturally occurring mitigating elements in

the area will continue their function and will not be severely damaged by those events (nor by stress produced by development activities). On the other hand, if there is a high probability of an event occurring that will substantially alter the form or function of the ecosystems of the area during the life of the management plan, then the plan should be modified accordingly. In the end, the plan may include at least a monitoring of changes produced by the event.

In any event, the assessment should include information concerning qualitative and quantitative aspects of vulnerability for the area within the protected reserve as well as for the adjacent areas that are affected by the hazard-mitigating effects of its structure and function.

Once the hazard and vulnerability assessments are completed and the information included in the basic planning and development documentation of the protected area, mitigation activities should be formulated as part of the management plan. These mitigation activities should include management of the ecosystems so that the naturally occurring mitigating elements are maintained to the greatest extent possible. The activities should also include the definition of other mitigation elements that may be put in place to augment the capacity of the natural ones, to supplement the expected loss of the natural elements, or to introduce hazard mitigation elements where none existed before.

These mitigation activities may be structural (research facilities designed and built to withstand a seismic event of a particular magnitude) or nonstructural (an alert system to warn visitors of flash floods or eminent volcanic eruption). They may modify the event itself (slope stabilization above an access road to avoid landslides), the hazard produced by an event (water storage to assure drinking water during a drought), or the vulnerability of proposed protected area activities (raising observation platforms above expected flood or storm surge elevation, given a particular event with a certain return period).

In the broadest sense of the term, mitigation includes the reduction of loss and suffering from a specific event. Therefore, management plans for protected areas should include emergency preparedness and disaster response actions, and should be prepared in full collaboration with those agencies responsible for implementation. This is necessary no matter how isolated, guarded, or "untouched" the area might be. Responders to a disaster interested in the survival rate of a particular species often arrive on the scene with, or before, responders dealing with human population needs. Preparedness and response plans should, first and foremost, consider the human population affected by the event, but should also consider basic scientific investigation interests. To those ends, plans should be built around the knowledge of what is vulnerable and why, and in anticipation as to where, when, and how an effective monitoring, alert, relief, rehabilitation, and reconstruction plan can be put in place.

The vulnerable populations, both human and otherwise, together with their habitats should be monitored and alert systems employed commensurate with the severity and frequency of the expected event. Accountability for visitors as well as staff is the primary concern. Appropriate response plans must be prepared and practiced where hazardous events would cause the need for, or hamper, rescue efforts. Even if no human population safety issues are part of the vulnerability situation, it is of interest to the management of the protected area to be able to anticipate and respond to emergencies, particularly if they do not reach the level of a declared disaster (which is understood to mean a situation beyond the coping capacity of the affected society).

Beyond the relief stage of a disaster, it is important that the protected area management plan consider what might be the range of possible rehabilitation and reconstruction responses suggested or thrust on the area. Understanding the possible and probable impacts of hazardous events permits review and, in some cases, decisions to be made before the event occurs as to what rehabilitation and reconstruction responses are appropriate and/or will be permitted. These responses, in and of themselves, can create further vulnerability or damage and can be counter to the sense and structure of the existing management plan for the area.

Conclusion

The creation, design, and management of protected areas are increasingly products of interdisciplinary efforts that reflect multiple purposes. And protected areas depend on multiple constituencies for their continuance and support. The emergency preparedness and disaster response community—with its national and international networks reflecting information, knowledge, and experience—should be brought into the partnership promoting protected areas. It is vitally concerned with the structure and function of the areas as ecosystems and as areas of human work and enjoyment. It can assist in promoting and defending the creation, design, and management of protected areas. Natural hazard management is part of protected area management.

References

Bacon, Peter R. 1989. *Assessment of the Economic Impacts of Hurricane Gilbert on Coastal and Marine Resources in Jamaica*. University of West Indies.

Bender, S.O. 1992. Disaster Prevention and Mitigation in Latin America and the Caribbean: Notes on the Decade of the 1990s. *Disasters and the Small Dwelling.* J. Aysan and I. Davis (eds.). London: James and James Science Publishers, Ltd.

Birkeland, Eldredge, and Grossenbaugh. 1983. *Interactions Between Tropical Coastal Ecosystems Mangrove, Seagrass and Coral.* University of Guam Marine Laboratory.

Brown, B.E., and L.S. Howard. 1985. Assessing the Effects of "Stress" on Reef Corals. *Adv. Mar. Biol.* 22:1–63.

Case, B., and M. Mayfield. 1990. Atlantic Hurricane Season of 1989. *Monthly Weather Review* 118 (5):1165–1177.

Clark, John R. 1988. *Report at the National Academy of Sciences Briefing on the Effects of Hurricane Gilbert in the Yucatan and Caribbean.* National Academy of Sciences, Washington, D.C.

Crow, T.R. 1980. A rainforest chronicle: a 30-year record of change in structure and composition at El Verde, Puerto Rico. *Biotropica* 12 (1):42–55.

Gable, F., J. Gentile, and D. Aubrey. 1989. *Global Environmental Change and Its Related Effects (sea-level rise) on the Coastal Caribbean.* 23rd Annual General Meeting of the Caribbean Conservation Association. 1–20.

Heyman, A.M. 1991. *Conservation of the Environment and Tourism: Impacts of Tourism and What Can Be Done to Reduce Them, with Emphasis on Coastal Caribbean Issues.* XVI Inter-American Travel Congress, 25–28 November 1991, Panama. Inter-American Travel Congresses Permanent Secretariat.

Lugo, A., E. Schmidt, and S. Brown. 1981. Tropical forests in the Caribbean. *Ambio* Vol. 10 (6):318–324.

McConchie, D.M. 1990. *Mangrove Forests as a Successful Engineering Alternative to Concrete Sea Walls in Coastal Protection Works.* University of New England, Northern Rivers.

National Academy of Sciences. 1991a. *Safer Future: Reducing Impacts of Natural Disasters.* National Academy of Sciences, Washington, D.C.

————. 1991b. *The Yellowstone Fire of 1988.* National Academy of Sciences, Washington, D.C.

Oaks, S.D., and M. Savage. 1991. *Natural and Anthropogenic Disturbance in the Yucatan, Mexico.* UCLA Latin American Center, Los Angeles.

O'Callaghan, Woodley, Aiken. 1988. *Project Proposal for the Development of Montego Bay National Park, Jamaica.* Prepared for OAS/DRDE. Montego Bay, Jamaica.

Organization of American States. 1991. Study of Integrated Tourist Circuits. *Integration of Itineraries of Tourism Attractions in Bordering Countries, Islands, and Sub-Regions, and the Enhancement of Individual Tourism Attractions.* XVI Inter-American Travel Congress, 25–28 November, 1991, Panama.

Rogers, C.S., L.N. McLain, and C.R. Tobias. 1990. *Effects of Hurricane Hugo (1989) on a Coral Reef in St. John.* USVI. 1–20.

Saenger, P., E.J. Hegerl, and J.D.S. Davie (eds.). 1983. *Global Status of Mangrove Ecosystems.* Gland, Switzerland: IUCN.

Salm, R. 1982. *Guidelines for the establishment of coral reef reserves in Indonesia.*

(FO/INS/78/061 [Special Report]). FAO/UNDP National Parks Development Project, Bogor, Indonesia.

Salm, R. 1984. *Marine and Coastal Protected Areas: A Guide for Planners and Managers.* Gland, Switzerland: IUCN.

Vermeiren, J.C. 1989. *Natural Disasters: Linking Economics and the Environment with a Vengeance.* Conference on Economics and the Environment, November 6–8, 1989, Barbados.

Woodley, J.D., *et al.* (19). 1981. Hurricane Allen's Impact on Jamaican Coral Reefs. *Science* 214:749–755.

Chapter 15

Protected Areas and the Tourism Industry

Stanley Selengut

Editor's Introduction:
This chapter presents a perspective on tourism and protected areas from the point of view of a tourist facility developer. Stanley Selengut has been a pioneer in developing "appropriate tourist technology" in some very sensitive protected areas. While many will argue that most tourism developments should be outside national parks rather than inside, the cases described here are examples of where the tourism facility helped in the further development of a more effective protected area. While the approach will not work everywhere, the working partnership between park management and tourism developer can provide income for the park, employment for local people, support for local culture, and enhancement of the natural environment. These kinds of tradeoffs may mean survival for some protected areas.

Introduction

This chapter describes partnerships between park managers and larger scale ecotourism, which can become the lifeblood of a protected area. First we will go on a tour through my own resorts, Maho Bay and Estate Concordia, in the U.S. Virgin Islands. Then we will travel around the world reviewing other successful examples.

Maho Bay and Estate Concordia

Sixteen years ago I signed a long-term lease for 5 hectares of commercially zoned private land within the U.S. Virgin Islands National Park. The site overlooks Maho Bay Beach, on the North Coast of St. John. It is a steep hillside property that rises 100 meters in just 600 meters. Below is a crescent of white sand beach fringed in places with coral reefs, supporting endangered turtles and abundant marine life. Before I began construction, the park super-intendent alerted me to the potential devastation that conventional building techniques might cause to the beach and reef. Were we to destroy the ground cover, or disrupt the site, the heavy seasonal rains would wash topsoil into the ocean, silting and smothering the coral.

I was a New York developer with little background in sensitive develop-ment. However, years ago I built resort housing adjacent to Fire Island National Park, a barrier island close to New York City. There, at a Sunken Forest, the Park Service had constructed elevated walkways to protect the rare vegetation from pedestrian traffic. I decided to build similar walkways to protect Maho Bay's plants and to avoid erosion. Our 5 m × 5 m "tent cot-tages" were also sited above the ground within the existing trees and plants. We connected these dwellings by the elevated stairs and walkways. We called the project Maho Bay Campground. The only permanent ground cover dis-turbance was the post holes for the columns that supported the walks and structures. Construction materials were wheeled along the walks and carried into place. Pipes and electrical cables were hidden under the walks rather than buried in trenches. The finished walkways flowed naturally through the trees and foliage. Guests walked safely from their tent cottages to the beach without trampling the ground. People fit comfortably into this natural setting. The 114-unit campground won the 1978 Environmental Protection Law Award, was featured in the *New York Times* Travel Section, and attracted the travel programs of major cultural and environmental groups. It seems we anticipated the growing market for "ecotourism."

Maho Bay is now one of the most profitable and highly occupied resorts in the Caribbean. In fact we are so popular that, long ago, we suspended most of our advertising programs. And after 15 years of operation and almost one million guest days, the property has been restored to a healthier wildlife and horticultural habitat than before it was developed. Recycling our water sup-ply accounts for much of this. We use and reuse over four million liters of fresh water each year for drinking, washing, and flushing. The treated water is then used to irrigate the hillside. From a distance you can hardly see the 120 structures hidden in the trees. This sensitive land use stimulated massive free publicity with attendant high occupancy.

The success of Maho Bay encouraged me to attempt the construction of a luxury resort with the same themes. I acquired a 10-ha commercially zoned parcel of waterfront property adjacent to the National Park on the south-eastern side of St. John. The property has many natural assets, beautiful views, a freshwater pond, a Danish ruin, and Nanny Point—a rocky promon-tory jutting out into the trade winds. All this is adjacent to Salt Pond Bay can imagine the marketability of a resort where walls open up to cactus and turpentine trees draped with orchids and air plants—and alive with parrots, iguanas, and parakeets.

Concordia should be like Maho, where we don't just preach conservation and caring. It's all around you at Maho Bay. Right from the start you get a feeling of communal cooperation. There is a "help yourself center" where you find free food and supplies left by departing guests. There is also Maho Bay's small but ample store stocked with healthful foods, biodegradable sundries and notions, and recyclable paper goods—all sensibly packaged, purchased in bulk, and devoid of plastics or fluorocarbons.

Our self-service outdoor restaurant offers healthful gourmet foods. A few simple phone calls assure us the shrimp we serve is caught in nets with turtle release devices, that the tuna we serve is not caught by nets that drown dolphins and other creatures. We always serve vegetarian alternatives to encourage people to eat lower on the food chain. Our kitchen uses biodegradable cleaning products, composts from leftover food, and employs water and energy saving devices. We also use boric acid instead of persistent pesticides for roach control.

The Pavilion is the center for interpretive functions. The Park Rangers give lectures on Wednesdays. On Tuesday, the watersports people discuss the fragile reefs and sea life and proper behavior while snorkeling, sailing, windsurfing, or scuba diving. Other days you may find concerts, dances, lectures, or other cultural activities to bring people together. But most activities are designed to promote health, fitness, and an appreciation of the world around us. A most valuable lesson that guests learn at Maho is how little one needs in life to be truly happy and comfortable.

You might ask a developer, "Why choose to build an eco-resort rather than a more traditional hospitality facility?" BECAUSE IT IS MUCH MORE PROFITABLE! The translucent wall fabric I use at the campground costs only $3 per meter, yet it is bright and durable. The sun shining through the trees patterns the walls. Inside it's like being in a Japanese shoji screen. What makes sense from an environmental and conservation point of view also saves money. Conserving power, water, and fuel is good business.

Native plant landscaping and feral animal control programs will lead to a heavily foliated, cooler landscape replete with indigenous wildlife and with little attendant maintenance. Recycling programs can also generate profits. A compactor allows the island to get enough aluminum in a container to make shipping cost-effective.

It is clear that it is more profitable to work with nature than against it, that environmental restoration can be a marketing tool, and that working with government and private agencies is better than working against them. In fact,

Beach, one of the finest white sand beaches in the Virgin Islands National Park.

The commercial zoning allows a density of 100 people per hectare. Were I to take advantage of this legal limit, I would surely face community and Park resistance. I presented preliminary development ideas at public meetings to find what would be acceptable to the people in the area, the local regulatory agencies, and the Park Service. They proved to be extremely helpful and far-sighted. We received assistance from the College of the Virgin Islands, which has an outstanding collection of Caribbean plants, and the Center for Marine Conservation, which could help in protecting the oceanfront and marine life, notably the endangered leatherback turtle.

Perhaps the most meaningful Park introduction was to the Society for Ecological Restoration. Their premise is that every parcel of land has a historical point where it reached its apex as a balanced ecosystem. Once this is identified, then there is a clear path to its restoration. In the Virgin Islands the apex of the habitat may have been in precolonial times, before Columbus landed in the "New World," before all the hardwood trees were cut and exported to Europe, before the land was farmed and grazed and heavy rains depleted the topsoil. Today the land I purchased at Nanny Point is a degraded forest with eroded topsoil. Alien species (many actually introduced) have replaced native plants and animals. Confronted with this condition, I was led to the theory of ecological restoration and a new development concept.

As you know, in conventional development the land is usually clear-cut and then re-landscaped with exotics such as grass and palm trees, thereby eliminating the land's value as a natural habitat. Instead, architect Claude Samton designed buildings with a small 6 m × 6 m footprint that could be placed between the trees. The units could then be connected by elevated pedestrian walkways. For each building, the land would be disturbed in only three places: a 6-m long trench on the high side of the slope and two point footings for columns on the low side. The decks would cantilever off the structure and open out above the vegetation.

Our goal is to leave as much valuable flora as possible and then to restore the habitat to its past glory by native plant landscaping. This means planting and irrigating vegetation and trees that have vanished from the site. Our plan emphasizes plants that will attract and support native birds and wildlife. At present there is a feral animal population of stray cats, wild donkeys, goats, and mongooses that the Danes imported to control the rat population. These have devastated indigenous plants and animals. With the help of park professionals we will try to regulate the number of these feral animals. We will try to reintroduce native wildlife, iguanas, parrots, and land-nesting birds. You many conservation groups such as the Audubon Society with large travel pro-grams might even become your customers.

As developers we usually do not stop to think about how we shape the future. Our plans and efforts have enormous environmental and social con-sequences. Environmental restoration is a new field. It is a form of develop-ment. And development can be a form of restoration. To achieve this demands a new concept of collaboration and commitment, a new partnership for conservation.

Traditional resort developers assess their customer's desires and make them comfortable in familiar surroundings by manipulating the environment. By contrast, ecotourist developers must enhance and develop the indigenous inventory such as wildlife, vegetation, traditional architecture, history, music, dance, and food. This offers their guests a learning experience as well as a cross-cultural exchange.

Maldives, Bali, and the U.S. National Park Service

The United Nations sent me on a consulting assignment to the Republic of the Maldives, a nation of 1200 tiny islands in the Indian Ocean southwest of Sri Lanka. The average height of each island is only 1–2 m above sea level. The total population of the Maldives is 200,000 people, of whom 50,000 live in the capital island, Male, and the balance occupy 200 inhabited islands throughout several atolls or groups of islands.

Transportation is mostly converted fishing boats transporting visitors from the only airport in Male to various resort islands. The islands are small, aver-aging only about 2 km at the widest point and spaced irregularly within their atoll. Most food is imported from Singapore or nearby Sri Lanka, except for seafood. The ocean is still crystal clear and bursting with sea life, an abun-dance of shell fish, reef fish, and the ever present tuna, which is the main food staple.

The government had the good sense to retain the ownership of these islands. They lease the uninhabited islands to resort operators for periods of up to 22 years. Most islands are surrounded by coral reefs creating a shallow lagoon of crystal clear water. These coral reefs protect the islands from flood-ing. The boats can only get to the outer reef. Long walkways span the lagoon from the dock to the island resort.

Tourism started about 10 years ago when the first resorts were built in the Male atoll. There were few guidelines at that time. Developers built what they believed would appeal to their visitors. Everything had to be imported, so most construction was done with available local material, such as coral, mined indiscriminately from the surrounding reefs, and local palm fronds for thatched roofs. With no enforceable guidelines, some development was commercial and unesthetic. Even the more sensitive developments featuring local architecture seemed to have a sense of imbalance. They were most proud of the massive electrical production equipment with banks of 250-kilowatt electric generators and state-of-the-art water desalination units to produce an abundance of fresh water. There was little attempt at conservation or rain collection. Large barges would pull in and unload diesel fuel into buried storage tanks. This seemed self-destructive in the country most threatened by global warming.

About five years ago, major erosion problems began through holes that the resorts had cut in the protective coral reefs. They tried to combat it with sand bags, then by constructing coral groins to build the beach using coral dug from the very reef that protected the islands. This caused more problems. So they then ringed entire islands with dug coral and filled the circle with sand dredged from the bottom, ruining their pristine marine environment and depleting the sea life. The concerned Maldive government finally developed a series of environmental guidelines. It outlawed spear fishing, surf casting, and taking any form of coral. It encouraged conservation and established new guidelines for resort development.

The first guideline forbids cutting down any tree without the express permission of the Ministry of Tourism. At Ari Atoll, the units were actually built with cutouts around the trees and trees touching the walls. No building could be higher than the tree tops. Each dwelling had to be separate and buildings could only cover 20 percent of the land area, leaving 80 percent with natural vegetation. Buildings had to be at least 5 meters behind the tree line at the shore, for a visual buffer.

The guidelines encourage native design and materials. However, the coral must be mined in designated areas that do not contribute to erosion. Composting is encouraged to build the soil. Wildlife protection laws are now in place and the birds enter the dining rooms to be fed. One resort protects the baby turtles until they are large enough to escape the birds and fish, then releases them. With these rigid development and operational mandates, the Maldives will still be unspoiled for our children—and their children.

The U.S. National Park Service hosted over 250 million visitors during the past year in its facilities. The NPS endeavors to have the architecture complement the nature and historic setting and express principles of rustic design through the use of local materials. The major strength of the U.S. National Park Service is its technical skills. The Denver Design Center employs over 750 architects, landscape architects, engineers, and naturalists. They have 75 years of experience in sewage treatment, garbage disposal, transporting visitors, creating water supplies and energy, and still preserving the integrity of a park.

In November 1991, more than 60 renowned architects, engineers, landscape architects, and naturalists met in the U.S. Virgin Islands. The purpose was to write the first handbook on Sustainable Resort Design, with guidelines for ecotourist development, which could be used by protected areas all over the world. They wrote on how to inventory local assets; how to build without destroying the environment; how to create energy, promote recycling, and treat waste; how to restore the site with native plant landscaping; on wildlife management; and—most important—how to insure the benefit to local people with jobs and business opportunities. The U.S. National Park Service will be publishing this handbook, which should be helpful to protected areas, lending institutions, regulatory agencies, and, in fact, anyone building anything on fragile natural property.

Conclusion

We have discussed examples of site adaptive construction in the Virgin Islands, effective environmental guidelines in the Maldives, and efforts toward sustainable development in the U.S. National Parks. But none of these is true ecotourism, because each addresses only one aspect of the whole. A true ecotourism facility within a protected area should become an integral part of the ecosystem itself. A working partnership with park management is the balancing force. Providing income for the park, employment for local people, support for local culture, and enhancement of the natural environment is a fair exchange for the privilege of renting some of the world's most precious real estate.

Equally important is the potential impact on visitors. By example and through interpretive programs the value of natural systems can be taught to millions of guests. Put very simply, nature makes the rules and if we follow them *everybody wins*.

Chapter 16

Energy Exploitation and Protected Areas

W.J. Syratt

Editor's Introduction:
Some people see oil exploration as an implacable enemy of protected areas. Yet it is also apparent that our modern society is dependent on high quantities and qualities of energy. Bill Syratt, senior ecologist at British Petroleum, one of the world's largest oil companies, calls for partnership built on a sensible balance that takes the needs and aspirations of people into account. He recognizes that the credibility of the international energy industry rests on its ability to deliver quality resources at reasonable cost while protecting the environment. He describes the responsibility of industry to eliminate or reduce impacts as far as is practical, including building restoration requirements into a project from the very beginning. But equally, the conservation movement must provide the necessary information that will ensure that the industry is well informed of conservation issues so that protected areas and the energy industries can work together to bring long-term benefits to all.

Introduction

This chapter presents an overview of the implications of energy resource exploration and exploitation for protected areas, with particular reference to oil and gas. It does not deal in detail with the use of that resource as a fuel, but identifies implications for protected areas arising as a result of exploratory activities.

Energy, or more specifically the resource from which the energy is derived, falls into four broad categories.

- Fossil fuels—coal, lignite, peat, oil, and gas
- Physical—hydro, tidal, geothermal, solar, wind, and wave
- Biomass—fuelwood, straw, dung, alcohol, biogenic gas, and so on
- Nuclear—fission and fusion

Again broadly speaking, the physical category is usually considered to represent renewable energy resources, as are many elements of the biomass category, particularly those components that are sustainably managed.

The full potential of the energy of the resource itself cannot be realized until it has undergone a transformation process to generate heat or electrical power. At the simplest level this may involve no more than a wood fire, the resource being used locally to provide the needs of a family unit. At the other extreme it may involve a complex chain of events that starts with the finding and winning of a resource, a transportation link to a processing plant (refinery, power station, or both), and a sophisticated distribution network, supplying the insatiable energy needs of the developed world (Table 16-1).

Energy derived from biomass resources is largely used close to its source of production to meet local needs. For the purposes of this chapter the biomass and nuclear categories will not be considered further other than to remind readers that there are environmental implications to both. In the former, for example, there may be forest clearance for fuelwood production and cane plantations. In the latter, mining of radioactive minerals for fuel takes place, and disposal of nuclear wastes remains a major problem.

Table 16-1. 1990 World Energy Statistics

Energy resource	World production (million tons)	World consumption (million tons oil equiv.)	
Oil[a]	3148.9	3101.4	(38.60%)
Natural gas[b]	1761.1 (mtn)	1738.1	(21.70%)
Coal: Bitumin+Hard	3121.6		
Coal: Lignite+Brown	1417.0		
Total mtn	2178.0	2192.1	(27.30%)
Nuclear	n/a	461.1	(5.7%)
Hydro	n/a	540.6	(6.7%)
All primary energy[c]	n/a	8033.3	(100%)

[a]Includes crude oil, shale oil, oil sands, and natural gas liquids—the liquids of natural gas recovered separately.

[b]Excludes gas flared or recycled.

[c]Economically traded fuels only. Excludes wood, peat, and animal and plant wastes that, though important in many countries, are unreliably documented in terms of consumption statistics.

Source: Attridge, 1991.

The Location of Energy Resources

Exploitable fossil fuel energy resources do not occur uniformly around the globe; they are concentrated in limited areas determined by geological processes and events that have taken place in the past. They have one thing in common—they were laid down in a sedimentary basin—but there the similarity ends. Today they may be exploited from many different locations ranging from mountainous areas to deep oceans, from the Arctic to the tropics. Fossil fuels occur "where you find them."

Physical energy resources are limited to where their intensity is greatest. Geothermal power can only be exploited from suitable hot spots; solar energy requires a high proportion of cloudless skies; wind-powered generators have to be located in windier districts, often coastal areas that may also be potential candidates for wave power; tidal power requires a large tidal range; hydro power requires a suitable head of water and a suitable water storage location (this in reality means that most hydro power schemes are restricted to upland or canyonland areas). To date only hydro power is widely exploited. Wind generator "farms" are increasing in number, for example in Denmark, but still represent a minor contribution to total world energy production, as do geothermal, tidal, solar, and wave power.

Energy and Protected Areas

Several categories of protected areas are recognized by the IUCN. It is probably a safe statement that around the world there are, for each category, examples that either already have energy resource exploitation taking place within their boundaries or within which the potential exists, whether that be a fossil fuel or physical resource. Potential environmental impacts exist in all cases. Though these will be site-specific and may be influenced by the type of protected area, they are broadly generic in nature and are greatly influenced by the type of resource being exploited (see below). Indeed, the very nature of a location that makes it attractive for energy exploitation, particularly for the physical resources, may be the same attributes that make it attractive as a protected area—for example landscape or coastal configuration.

The winning of an energy resource is only one part of a larger picture that involves the economic delivery of the resource to the user. This involves transportation and conversion. The conversion of the resource to energy itself or some other energy resource (such as gasoline and fuel oil, or chemical feedstock) has implications for protected areas. In the case of physical resources the conversion occurs at the point of exploitation, and the energy is used locally or exported (in some cases internationally) as electricity. The only exception is geothermal energy, which may be transported over limited distances and used for heating. In the future the electricity from these resources may be converted to transportable materials such as hydrogen.

Fossil fuels may be used locally, at or close to their point of extraction, or exported. Lignite (brown coal) and peat contain large quantities of water (up to 80 percent by weight). This factor alone limits the value of lignite and peat as energy resources tradeable on a global basis. Most is converted to electricity in "minemouth" power stations. Electricity transmission has its own aesthetic impacts—pylons across an otherwise unspoiled landscape.

Oil, gas, and coal are widely traded around the world. At present, they are the only easily transportable and concentrated energy resources and may be used far from their area of production. The transportation of oil in particular poses a risk to protected areas from accidental losses from ships or pipelines that pass close to protected areas. Losses of this sort are unpredictable—their location cannot be pin-pointed in advance. Oil spill contingency plans are needed for risks of this nature.

Similarly, risks arise at the point of conversion. Most refineries around the world, and many power stations and chemical plants, are located on major estuaries and rivers. Many of these locations have protected wetland areas of international importance close by, occasionally adjacent. Quite apart from the localized impacts that effluent discharges may have on a protected area, there is also a risk of feedstock or product spill from the conversion plant, or a transfer terminal associated with it, affecting a wider area. Again, it is necessary to consider protected areas in a contingency plan, but because the location of the potential spill is more predictable, it is possible in some cases to take more positive protection measures, for example the installation of permanent boom mooring positions.

While coal and gas losses are not unknown, should they be lost their impact on the environment is comparatively small. Coal is practically inert, but will have a localized smothering effect, while gas spills usually disperse rapidly. The latter pose a real risk of explosion, probably of greater concern to the human environment than to the natural one, and methane contributes to the "greenhouse effect."

During the conversion process various wastes arise. Though there is an ever-increasing drive to reduce these as much as possible, their total elimination is currently not practical. This is particularly true for gaseous emissions from fossil fuel combustion. All fossil fuel combustion will give rise to carbon dioxide and water vapor, but there will also be varying concentrations of other gases. Emissions of these will depend on fuel quality and combustion efficiency. The former is particularly important for sulfur dioxide emissions;

the latter is particularly important for nitrogen oxide emissions. Both contribute to acid rain. This may affect areas remote from its source, including sensitive upland areas and freshwater systems, some of which may be protected areas. There are, however, technologies available for controlling both sulfur dioxide and nitrogen oxides.

The Value of Protected Areas to Industry

Natural ecosystems are geared to the particular environmental characteristics of the area and are able to respond to natural inputs and control outputs. Many of these attributes are of direct value to industry, especially where extraction of fossil fuel energy resources is concerned. The area may provide flood control, erosion control, a water resource, construction materials, recreation, and, occasionally, food. No development takes place without the utilization of at least one of these attributes, frequently more. In developed areas it may be possible to obtain needed support from already developed sources, such as water supply, timber, and so on. In more remote areas this is not possible, and exploitation of the natural resources of the area becomes more important.

It is therefore in the long-term interests of project developers to maintain the natural system as intact as possible. The costs of a project can be significantly reduced if flood and erosion control can be avoided by judicious project planning, leaving the bulk to be achieved by the natural system. A catchment area that acts as a natural reservoir, supplying sustainable quantities of high-quality water, can substantially reduce water treatment or water transportation costs. The use of natural construction materials can reduce costs by reducing the amount of materials to be brought in, especially in the early stages of a project when materials import is often a time-consuming activity before supply routes have been set up.

Seismic exploration, for example, often makes use of a high proportion of natural materials: timber for camp construction; felled polewood for decking, walkways, and bridging; and fruits to supplement rations. In this respect, the presence of a protected area nearby can offer those resources. The quantities used will, in most cases, be small and sustainable, the disturbance minimal, and the overall impact well within the capability of the system to absorb. In some cases overall impacts may be reduced by utilizing local resources; for example, in remote areas if all construction materials have to be brought in, a larger infrastructure has to be in place to handle it.

During the operation of a project the existence of a protected area close to or even surrounding the project area can offer continued erosion control, flood prevention, and water supply. On Bintan Island, Indonesia, which has been largely deforested, there is still at least one fragment of high-quality tropical rainforest that continues to protect the catchment that supplies the drinking and process water to an oil terminal.

More project developers are beginning to realize the value of a buffer zone around their project areas. Control over this zone enables discouragement of opportunist settlers who often set up camp at a boundary fence hoping to benefit from the project, creating a nuisance or health hazard. Such a zone also offers potential for conservation. In the Jamari National Forest Reserve, Rondonia, Brazil, a tin extraction operation that occupied less than 3 percent of the area of the reserve undertook the protection of the rest of the reserve and provided security against illegal logging and mining activities.

More recently we are seeing the development of the "carbon reserve" principle, where promoters of energy projects in the developed world are being obliged by their government to either purchase forests in less developed countries for long-term protection or plant new forests for the purposes of absorbing the carbon dioxide produced by that project. This principle is well advanced in the Netherlands.

Should this trend continue, it is possible to envisage a day when pristine areas of forest will be sought as a resource not for the value of what can be gained from them, but for the sole purpose of providing a carbon sink. If that stage is reached it is probable that funding for protected areas will be based not on altruistic reasoning, where it is difficult to quantify an economic value, but on sound commercial reasoning, where the cost of the demise of such an area will be clearly understood by society, and where the cost of protecting that area is borne by society as the consumer of energy.

Partnerships for the Future

The future of protected areas is already in the hands of society at large. Unfortunately there is no consensus within society as to how protected areas should be managed. At one extreme is the element that demands absolute protection at all costs; at the other extreme is that element that wishes to see "progress" at whatever cost to the environment. What is needed is a sensible balance, in particular a balance that takes the needs and aspirations of people into account, wherever those people are. It is understandable that those who "have" wish to continue to have and those who "have not" should aspire to have. Uncontrolled exploitation of energy resources will not necessarily achieve the aspirations of either group.

The energy industry has long been counseling the need for energy efficiency and energy conservation. The international energy market demands high-quality energy resources; developing countries are demanding that energy be made available to them at reasonable cost; society is beginning to demand that those resources be won with minimal impact on the environment. The credibility of the international energy industry rests on its ability to do just that: deliver quality resources at reasonable cost while protecting the environment. This is a tall order but not impossible if the parties involved are prepared to cooperate.

One way of achieving balance for a protected area is to consider its value not only in terms of its strict conservation but also in terms of its worth to the government of the country as well as local people. Within this framework it will be possible to identify a wide range of values; at one extreme the greatest value may be immediate clearing and settlement with no value attached to conservation; at the other extreme, that it should be protected at all costs—the polarization that was identified earlier. Only in rare exceptions should either extreme be necessary. It may be that people pressures are so great that conservation is, genuinely, a luxury. Likewise there are probably cases that deserve conservation at all cost. In the case of the latter, the value to the government and local people can only be realized if the international society pays for that protected area; in other words, its value is that of a long-term lease income.

Greater value may accrue to both the country and conservation if sustainable development can be achieved from the area, through such means as managed timber extraction, agroforestry, or natural product harvesting. Even greater benefits may be obtained if there is a resource within the protected area that can be developed with minimum disturbance to the protected area. An obvious example would be oil and gas exploitation. The land loss involved is small but the value of the resource being extracted is very high, often far higher than that from a timber extraction program or a settlement program. In terms of a benefit to the government and peoples of the area, there would be employment and an immediate high return that could be invested in sustainable developments for the future. In terms of benefit for conservation, the unit area value would be produced from a small proportion of the total. Part of the exploitation license could be an obligation on the part of the exploiter to provide resources for the remainder of the protected area—a "good neighbor" approach like that adopted by Kaltim Prima Coal toward Kutai National Park, a policing agreement like the tin operations in the Jamari National Forest Reserve, or sponsorship of direct conservation measures.

Partnership goes beyond the "you scratch my back, I'll scratch yours" type of approach. True partnership involves an understanding of each other's point of view, each other's aims, what can be done and what cannot. It involves the drawing up of codes of practice that allows both industry and the conservation movement to approach a potential development from a point of common understanding. Recently the E and P Forum (1991) and the IUCN (1991) separately published guidelines for oil and gas exploration in tropical rainforests. The two bodies have agreed to cooperate on the production of future guidelines, for example, for cold regions.

Conclusion

Above all, partnership involves working together. It places an onus on the energy industry to find out about the area in which it has interests and to undertake good environmental protection management, especially where sensitive and protected areas are concerned. It places an onus on the industry to eliminate or reduce impacts as far as is practical, for example, through access control; minimization of the project "footprint," as is currently being practiced by the industry in Alaska; strict control of wastes and pollutants; and restoration requirements to be addressed at the front end of the project, not as an afterthought. But equally it places an onus on the conservation movement to respond in a positive way, to provide much-needed information, to ensure that the industry is advised of the issues, and to work together to bring long-term benefits to all.

References

Attridge, G. 1991. *BP Statistical Review of World Energy*. The British Petroleum Company p.l.c., London. 37 pp.

E and P Forum. 1991. Oil Industry Operating Guideline for Tropical Rainforests. Report No 2.49/170, *E and P Forum*. London. 17 pp.

IUCN 1991. *Oil Exploration in the Tropics—Guidelines for Environmental Protection*. Gland, Switzerland: IUCN—The World Conservation Union. 30 pp.

Roberts, T.M., and L.W. Blank. 1990. Energy Exploitation and the Environment in "Renewable Energy." In the Proceedings of a Symposium held by The Institute of Biology, London. 1990.

Chapter 17

Protected Areas as Investments

Mohamed T. El-Ashry

Editor's Introduction:
Investments in protected areas are becoming an increasingly important, though still small, part of the investment portfolios of the major development assistance agencies. This chapter, written by the director of the World Bank's Environment Department, provides an insight into how the Bank is seeking to foster productive partnerships for expanding investment in protected areas. Drawing on advice from the scientific and academic communities, the Bank is seeking to identify problems and define issues in response to which the Bank will assist member governments to devise operational policies and programs and to mobilize additional financial resources. It is apparent that the Bank will find such investments most appropriate when they link protected areas with surrounding lands, address both the fundamental causes and the symptoms of weak protected area management, and see development and conservation goals as complementary and mutually reinforcing. Protected area agencies seeking partnerships with the World Bank will need to ensure that the finance ministries are in support of such partnerships. This process will often be facilitated by pressure from NGOs. The principles for promoting effective partnerships between development banks and protected areas are relevant far beyond the World Bank.

Introduction: The Heightened Environmental Consciousness

Reflecting on this congress immediately brings to mind the tremendous changes that have occurred since the previous such meeting in Bali a decade ago. More recent initiatives at the international level range from the 1987 Brundtland Report, which elevated the whole level of concern over environmental issues, to the 1992 UNCED meeting in Rio. As a result of this increased consciousness, environmental issues are now routinely on the agenda at meetings of heads of states and ministers, including the Group of Seven (the leading industrial nations). It has proved possible to negotiate the

Montreal Protocol on ozone depleting substances and to support this initiative with the Interim Multilateral Fund (IMF)—now supplemented by the Global Environment Facility (GEF) of over US$ 1 billion. The GEF covers a wider array of worldwide environmental concerns: global warming, pollution of international waters, and biodiversity protection. Of potentially even greater significance to us all is the proposed Convention on Biological Diversity.

At the national level, another most favorable sign is the fact that developing country decision makers generally share the environmental concerns of their counterparts in the industrialized nations and agree that environmental degradation has become a significant barrier to development. At the same time, the developing world faces the dilemma of simultaneously coping not only with urgent issues like malnutrition and poverty, but also with the more recently recognized problems of the environment. It is heartening that, generally, there is significant scope for reconciling development goals with responsible stewardship of the environment. Finding specific programs and policies that emphasize this complementarity is an important challenge for us all.

Finally, if I may be permitted to reflect on my own institution, one could point out that the World Bank's very presence here and our wholehearted support for this congress are themselves a reflection of the extent of the movement that has occurred. In fact, it is only when one prepares for a gathering such as this that one realizes the remarkable changes within the Bank during the past five years, from the environmental perspective. Most obvious are the structural changes, following the creation of the central Environment Department in 1987, together with an Environment Division in each of the operational regions. On the output side, a significant share of projects now have major environmental content and a mandatory process has been established for environmental review of all projects and assessments as appropriate. In addition to project lending, a portfolio of activities supported under both the Montreal Protocol and the GEF has been established, and a dialogue has been initiated with most countries on environmental issues, leading to preparation of National Environmental Action Plans. All of this has required a substantial increase of staff focusing on environmental issues. Prior to the reorganization in 1987, there was only a small unit of some half dozen persons; now there are at least 140 environmental staff, of whom about 30 are actively engaged on biodiversity-related work.

The greatest changes in the Bank, however, have been in several ways that may not be apparent from outside. First, the management and staff at all levels are now interested in, and responsive to, environmental concerns. Second, the additional requirements for environmental assessment that have been introduced have meant that the range of issues addressed in economic and project work has been widened. Environmental concerns are being more systematically incorporated into all Bank operations, but the full effects will be seen only in the late 1990s as these projects come to fruition. Third, on the more formal level, this is also reflected in the many policies and guidelines that the Bank has produced. Of particular relevance are the Forest Policy Paper prepared last year (which had a strong environmental focus), the World Development Report this year (which deals with the environment and development), and the directives and guidelines on environmental assessments and wildlands, as well as others in preparation.

The environmental movement has resulted worldwide in an increased emphasis on participation at all levels and greater openness and access to information. These trends have also influenced the Bank, and led us into a wider range of cooperative activities than ever before, basically at two levels. One has been the collaboration with individuals and groups outside the "traditional development community"—particularly those in the environmental NGOs and academia—in relation to both research and operational activities. The second has been a significant increase in the extent to which the Bank has responded (often together with other regional or international organizations) to requests for assistance in the development of mechanisms and/or programs needed to tackle transnational environmental problems. Examples include the evolving programs for the Amazon rainforest or the rehabilitation of the Mediterranean, Black, and Baltic seas. Thus, the Bank is being drawn into an increasing range of innovative partnerships, which may help form the basis for helping countries tackle some of the major environmental issues confronting them.

In this brief chapter, I would like to consider more specifically the issue of the future of protected areas within the context of this evolving framework.

The Changing Role of Protected Area Management Agencies

The variety and types of pressures on protected areas and protected area management institutions have increased dramatically in the past few decades, and are likely to continue to do so in the future. Population growth and a continuation of traditional resource-intensive consumption patterns are two of the critical factors leading to the destruction of natural habitats. While encroachment into protected areas or illegal extraction of wildlands products is often blamed on the poor, they often represent the proximate rather than the root cause. In many cases, sheer poverty combined with inequitable land

and resource distribution leave the poor with few alternatives. In other situations, protected areas have been adversely affected by national-level policies
that provide incentives for improper resource use. Examples are inadequate
pricing policies for timber extraction, subsidies, or tax credits for inappropriate land uses, and road construction and settlement in areas unsuitable for
agriculture.

In response to these multiple threats, protected area management agencies
are being asked to deal with larger areas, many other government institutions,
and new issues. They are increasingly called on to work at the national level
to identify and promote policy reforms that will diminish the threats to protected areas. They are also being asked to work closely with other agencies to
develop economic alternatives for the rural poor living adjacent to protected
areas. Protected area management agencies are also being asked to coordinate the activities proposed by the scientific, NGO, and international donor
communities. At the same time, these agencies are expected to maintain their
more traditional roles, emphasizing strong management that balances effective enforcement, park maintenance, community extension, tourism, and
research needs.

The pressures on protected areas, combined with the variety of roles that
protected area management agencies are being asked to perform, often place
overwhelming demands on them. Therefore, it is essential to realize that adequate levels of funding are only one of the many components needed to
strengthen protected area management. Strong legislative mandates, policy
reforms, and good interagency and external coordination are as vital to protected area management as a well-trained and equipped staff. In addition this
staff must increasingly be reoriented toward the more cooperative philosophy
required, by training them in the necessary communication and "people"
skills, which will enable them to draw on local communities more effectively,
especially to enhance women's participation.

In other words, short-run financial constraints may not be the only limiting factor to park protection in many countries, especially given the current
"boom" of interest in biodiversity conservation (although this might not last
at its present level of intensity). Of greater significance is the lack of capacity
of governments and local institutions to effectively absorb and manage large
infusions of funding—this itself being a manifestation of past financial constraints. In fact the most urgent need may be for modest but very reliable and
consistent levels of funding over the longer term, so that institutions can systematically improve their capacity for action, as opposed to fluctuating wildly
up and down depending on whether there happens to be a flush of external
funding. The main challenge for many of us will be to ensure that such modest levels of funding are sustained and used effectively over the long term.

National Responses to Changing Conservation Priorities

The recognition given to the importance of biodiversity conservation, and the key role that protected areas have in biodiversity conservation, has never been greater. A decade ago, issues such as biodiversity and protected areas rarely would have been part of either national or global agendas. But priorities at the national level are changing and an array of responses that will help improve protected area management, even in the face of shrinking national budgets, are emerging.

Many countries are starting to link environmental concerns, including issues related to protected area management, to national development plans. Unsustainable and ill-managed growth are often as responsible as poverty-related factors in explaining the misuse of resources. New methods of analysis that capture the costs of resource depletion, such as national income accounting, demonstrate that resource degradation does have both immediate and long-term costs. Once these techniques are more routinely applied and accepted, they will provide an incentive to policymakers to formulate national development plans that do not sacrifice the natural resource base to achieve short-term economic growth. Although countries are beginning to use these methods to place a better value on their environmental assets, considerably more is needed to demonstrate the value of wildlands and wildlife resources to national economic decision makers.

There is increasing recognition that, even using this broader understanding of the role of natural resources, it will only be possible to allocate a limited share of the biologically critical areas for strict preservation purposes. We must also recognize the increasing importance of the need for partnership between people and nature outside protected areas—for example, the management of buffer zones and multiple-use areas where exploitation of resources at sustainable levels is permitted, as in forest reserves and indeed within some national parks themselves.

Within countries, much can be done at relatively little cost. Many of the factors that threaten existing protected areas and make management of these areas more difficult (such as deforestation) can be dealt with effectively through policy reforms. These reforms are often based as much on sound economic criteria as they are on ecological criteria. For example, appropriate price policies in granting forest concessions or for fuelwood can have a substantial impact on deforestation. Tax incentives and policies, subsidies, credit, and land distribution can all exert a powerful influence on deforestation. In response, increased attention has been given to identifying such policies through National Environmental Action Plans and similar exercises. Such exercises will help to identify needed reforms within countries, many of which can be accomplished with very low levels of financing. At the same time it has to be recognized that such policy reforms still need to be backed up by effective implementation.

Another positive response to decreasing national budgets has been the range of innovative financing mechanisms to pay the costs of protected area management. The best known of these mechanisms have been debt-for-nature swaps and the establishment of trust funds and endowments. For example, the Bhutan Trust Fund has been established with GEF financing as a guaranteed source of funds for conservation initiatives in the country. Annual interest from the fund will be used for training natural resource professionals, institutional support, forest and wildlife surveys, environmental awareness programs, and protected area system planning and management. This fund is just one example of creative financing arrangements being established in many countries. However, caution must be exercised to insure that countries do not make conservation an "off-budget" item, relying primarily on revenues from external sources or from sources such as ecotourism. It is vital that conservation and protected area management remain well integrated into national priorities, and vice versa.

In summary, three important factors at national levels may offset financial constraints. The first is the increased attention being given to environmental issues at national levels. While actions may not yet match rhetoric, the enhanced visibility of the environment is important. Second is the potentially widespread impact of policy reforms in improving environmental management and mitigating some of the causes that lead to the destruction of natural habitats and protected areas. Efforts are underway to identify and implement appropriate policy reforms in many countries. Finally, countries are devising new ways of externally financing conservation, in cooperation with the international community.

International Aspects of Biodiversity Conservation

Biodiversity preservation has also been recognized as an issue of global importance. It has been internationally accepted as a priority akin to other pressing global issues, such as global climate change and deterioration of the ozone layer. Because of their role in biodiversity conservation, protected areas have received increased attention as well. Until recently, protected areas were viewed as primarily benefiting the countries in which they were located. But we have come to realize the regional and global importance of protected areas and their links to the global environment.

The regional importance of protected areas has become especially evident in the case of transboundary issues, which demonstrate that adverse actions in one country can have direct effects in another. Watershed protection is perhaps the best known example of conservation in one area providing sub-stantial benefits in other countries.

Increasing recognition that the quality of the global environment can sig-nificantly affect localized areas has two important implications. First, it sug-gests that all countries have an implicit stake in what happens in any other country. Second, it demonstrates the importance of global actions to address what were once thought of as localized issues, and conversely it underlines the inadequacy of relying solely on local protected areas.

These revelations have led to discussions about the appropriate distribu-tion of the costs associated with protected area management. Local residents often are least able to bear the costs imposed by protected areas. What is even more unfair is that they should bear these costs while benefits are distributed nationally or globally. The willingness of the international community to con-tribute the financial resources needed for protected area management is in part a result of the recognition that the costs of biodiversity conservation can-not be borne disproportionately by poorer people and countries, when the benefits accrue globally.

A number of initiatives have arisen to help governments decide how funds can be more equitably distributed. For example, the *Global Biodiversity Strat-egy*, a product of the World Resources Institute, IUCN, and UNEP, has broad support because it was developed in consultation with numerous govern-mental and nongovernmental agencies worldwide. The Bank agrees with the actions identified in the report to conserve biodiversity—many of which are related to protected areas. A number of other worldwide efforts are ongoing as well, including the second World Conservation Strategy (*Caring for the Earth*), UNCED's Agenda 21, and the Tropical Forestry Action Plans. Fur-ther regional efforts, such as the "Our Own Agenda" initiative, a joint effort of the Inter-American Development Bank and Latin American and Caribbean countries, are also underway.

These diverse efforts highlight the attention being given to biodiversity conservation and protected area management in many quarters. Yet they also highlight the lack of consensus on what the priorities are and how much it will cost to implement the recommendations. The estimates vary substan-tially, from US$ 52 billion in *Caring for the Earth*, to approximately US$ 30 billion in Agenda 21, to US$ 17 billion per year in the *Global Biodiversity Strategy*. While the pursuit of greater funding for biodiversity conservation goes on (much of it focused on protected areas rather than *ex situ*), it is essen-tial that some consensus be reached concurrently on urgent national and global priorities, and that implementation efforts are well coordinated. In addition to helping frame the agenda of priorities in biodiversity conserva-tion, the Bank is particularly well placed to systematically facilitate practical implementation. We are willing and able to be a responsive bridge between the scientific/NGO community and economic–financial decision makers. This is an important element of the partnership I alluded to earlier.

International development institutions have given higher priority to protected area management, reflecting the increasing belief that development must be linked to conservation. This shift in the priorities of international development agencies is reflected in the significant changes that have taken place at the World Bank. For example, the World Bank's Asia Region Biodiversity Strategy recognizes that the conservation of biodiversity involves many types of activities, but it argues that a focus on protected areas is the best place to start. Funding for at least one major initiative for biodiversity conservation in each country within Asia appears likely. In the Africa region, the Bank is probably the single most significant source of funds for biodiversity conservation. For example, the proposed Kenya Protected Areas and Wildlife Management project amounts to $150 million of which $60 million is from IDA—the soft loan facility of the Bank. The Bank is likely to continue as a major source of funds for enhanced protected area management in all regions.

The need for collaboration in promoting protected area management is becoming one of the most important emerging messages. Partnerships among NGOs, governments, and international development institutions are taking place at many levels, from international-level cooperation to financing large funding transfers such as trust funds at national levels, to implementing projects such as the Integrated Conservation Development Projects (ICDPs) that primarily affect protected area management at local levels. Agencies involved in international conservation are adopting different types of roles, depending on the activity. As a result, these partnerships are leading to new management structures, funding mechanisms, and financial and technical support for conservation.

Partnerships draw on the strengths of each of the participating agencies in ways that enhance biodiversity protection. Increasing examples of collaboration, with different partners assuming different roles, from financing to research to management and project implementation can be expected in the future. This interest will bring increased financing for global biodiversity, as well as an array of technical, managerial, and institutional capabilities.

Conclusion

Recognition of the vital role of protected area management has increased dramatically in recent years. Yet the awareness of the important function of protected areas in conserving natural habitats and protecting biodiversity is only a preliminary step. It will be necessary for the agencies responsible for protected area management to move beyond their traditional roles and into new roles that stress advocacy, policy reform, and interagency coordination. This role is becoming increasingly important for many protected area management agencies, yet it is one that cannot happen overnight. At the same time agencies in other economic sectors (especially the intrastructural ones like energy and transport) must be persuaded to incorporate environmental and biodiversity concerns more effectively into their projects and policies.

I reiterate that while it will be possible to allocate a share of the biologically critical areas as strict reserves, the majority of the surface of the earth and its resources will be subject to use for human activities or multiple purposes and that therefore the majority of its flora and fauna will have to coexist with human beings. The challenge will be to evolve patterns of use that enable as many as possible of these multiple demands to be met, at least in part, and to establish the incentive policies and institutional and legal frameworks to enable this to happen. The *Global Biodiversity Strategy* is particularly valuable here, in providing a basis for continued dialogue toward this end.

This brings me back to the issue of partnerships that I raised at the beginning. Considerable progress is being made in the recognition that economic development can be pursued in parallel with conservation of biodiversity. This complementarity applies to ways in which conservation as a principal objective can be pursued while, simultaneously, other benefits are derived mainly for the local populations. It also applies to methods of identifying the broader environmental impacts of particular patterns of development, and in devising means of mitigating them. At the same time it has to be recognized that measures that do not meet the aspirations of people will not succeed. In the end human needs and aspirations will be paramount, and therefore those pursuing conservation goals should recognize and harness these other motivators of human behavior.

Chapter 18

How to Involve Women in Protected Area Issues

Beti Astolfi

Editor's Introduction:
Beti Astolfi is the senior advisor to the United Nations Development Fund for Women (UNIFEM), and as such is very well placed to advise on how a previously unappreciated half of human society—namely, women—can be brought into more active partnerships with protected areas. It is clear that women have had a major impact on resource management, but scientific research, protected area management, and rural development have all tended to focus on the male part of the agenda. Astolfi argues that new and specific policies are required to enable women to reach their full potential as partners in protected areas. These include appropriate research, training in gender issues for protected area staff, new policies, and wider communication on issues of women and protected areas.

In his keynote address at the Bali Congress in 1982, Marcos Flores Rodas, Assistant Director-General of FAO, appealed to participants "to focus attention on the interest of rural people in developing countries who live in the vicinity of national parks and protected areas." We can happily note that these words have not gone unheard.

One of the main goals since the Bali Congress has been to take the needs of rural communities into account at all stages of planning, establishing, and maintaining protected areas. In doing so, was it genuinely assumed that involving rural communities necessarily included all of its members, women as well as men? Was it assumed that tapping the knowledge of rural populations for the purposes of conservation necessarily applied to all social groups? Was it also assumed that evaluating the needs of rural populations and their access to natural resources was done from the vantage points of both men and women? I raise these questions because, unfortunately, even when people are on the agenda, somehow women remain invisible.

The mere phrase "Women and Natural Resource Management" tends to raise eyebrows more than consciousness. Why is this so?

One reason is that development planners are still guided by a monotony of misconceptions about the contributions of women to national development. For example, at agricultural extension and policy levels, the image of the male farmer still prevails. Despite the fact that an FAO study in Africa found that women make up 80 percent of food producers in some countries, fewer than 10 percent of these farmers received benefits from extension services.

Another reason is that development practitioners are only beginning to understand the different, but interdependent, roles of men and women and how these roles have a direct impact on their relationships with forest and land resources. We need to better understand gender roles in order to correct misconceptions about them. The process for doing this builds on a very simple premise: Many differences between men and women are socially constructed and can be changed. What kind of vision can we have for preserving our remaining protected areas unless we are willing to change our perception of human reality?

A further reason is that conventional national accounting includes women's economic contributions to society only if they fall under the formal definition of "productive work force." This creates serious discrepancies in national accounts and leads to inadequate policy decisions. As we have witnessed time and time again, when women's tasks, responsibilities, and contributions are ignored, the results we envision elude us.

Joan Martin Brown, Director of the UNEP Washington office, has rightly pointed out that the contribution of women to maintaining ecosystems is, in reality, a "shadow subsidy." Without these contributions by women, national economies would be hard pressed to preserve a large proportion of the natural resource base. Two-thirds of the world's poor are women who face the challenge of providing water and fuel, food, and fodder every single day. Faced with ever-dwindling supplies, these women have a powerful incentive to protect the environment. They know all too well that walking that extra mile for fuel means they have less time for more productive activities.

At a recent meeting on Women and Biodiversity, Walter Reid, from the World Resources Institute said: "Four or five years ago, nobody gave indigenous peoples much thought, but now you can't talk about forests without talking about them. People concerned about saving biodiversity should demand the same kind of visibility for rural women."

The section on the Role of Women in Protecting Biodiversity contained in the *Global Biodiversity Strategy*, launched at the IVth World Congress on

National Parks and Protected Areas, is a testimony to 10 years of changing attitudes since Bali, and the efforts of many to put women's issues firmly on the global agenda. As leaders, however, our duty is to look ahead, and our resolve should be that we cannot wait another 10 years to make women a recognized resource. A compelling reason is that "nature will not wait that long."

Let us look at some of the specific contributions women have already made. Women have developed a thorough knowledge of plant growth, maturation, and reproduction. It was through women's interventions that early grain seeds such as sorghum and wheat were culled and valued. It was through their observation and experimentation that ash was used as fertilizer and that mulching, terracing, fallowing, and crop rotation were established as fundamental elements of sustainable agriculture. Studies of hill and mountain economies in northern India demonstrate that ecological stability through the sound management of forest, agriculture, and livestock is largely dependent on rural women's knowledge and labor in maintaining biomass flows. When monoculture crops displace indigenous varieties, however, the quality of biomass for fodder and fertilizer is greatly reduced, resulting in declining soil fertility and decreased food production. The Center for Science and Environment in India has even called for a redefinition of poverty, "not as a shortage of cash but as a shortage of biomass."

Despite these truths, women's potential to be part of a more permanent solution to maintaining biodiversity is neither fully understood nor taken into account. Biases have been inherent in scientific research, in protected area management, and in agrarian and rural development approaches. These biases have led to an inadequate analysis of women's productive roles and have resulted in undervaluing their contributions. For the past 15 years, UNIFEM has worked to ensure that more people recognize that women are an invisible force producing the visible goods. It is our role as a catalyst within the United Nations system to mobilize resources for women's economic activities and to give voice and visibility to the enormous contributions of women.

In this, we are not alone. In preparation for the UN Conference on Environment and Development, a number of international events have been organized that have given visibility to rural women and their role in preserving natural resources. One such event, the Global Assembly on Women and the Environment, sponsored by UNEP, brought together about two hundred success stories of women, from all over the world. These incredible women have in one way or another succeeded in protecting ecosystems against all odds and with little means. Case after case demonstrated where women have initiated activities and created support systems for community survival without any further degradation of the resource base. These case studies are illuminating examples of how women all over the world are quietly practicing what so many people preach: that preserving our planet requires commitment and action from each of us.

When 15 women in Bamako, Mali, armed with little more than basic training, scant funds, and two trucks, can build a successful sanitation service for 18 thousand urban poor—*that* is commitment. When the government of Yemen adopts a national policy to disseminate bio-gas technology pioneered by women through a UNIFEM project—*that* is both exciting and powerful.

UNIFEM's work was founded on a belief in women's capacity for leadership, for taking control of their own lives, and for having a positive influence over their families, communities, and nations. Over the past 15 years we have contributed to an international effort to integrate women into mainstream development. However, a recent publication, *The World's Women: 1970– 1990*, points to gaps in information and barriers that women still face. It especially argues for the need to have specific national policies to enable women to rise to their full potential in social, economic, and political life.

Only if policies are developed to promote community-based research, to empower women farmers, to educate and train women for careers in science, in agriculture, in forestry, and in protected area management, will we move from rhetoric and resolutions to results.

To close the gaps in our knowledge with regard to the role of women in protecting natural resources and biodiversity, it is important that an emphasis be placed on data gathering. Relationships between human populations and protected areas vary according to ecosystems, cultures, and production systems. Unfortunately, documentation on women's interaction with protected areas is very scarce. Specific indicators based on women's interactions with the environment could be developed and monitored as part of an early warning system for environmental stress. It is especially important that park project teams be provided with gender training prior to information gathering, as sensitivity to women's issues and cultural situations impacts on the relevance of data collection.

While we know that women gather grasses long before dawn in the national parks of Pakistan and that they seek out fuelwood in protected areas throughout the Sahel, we need further study on subsistence strategies undertaken by women if protected area management is to fully reflect the needs of local populations. This process of information gathering must be done in the field, within communities living inside and around protected areas. Uses of natural resources by women must be listed, for each ecosystem. Women's efforts to preserve or restore natural resources must be carefully investigated and analyzed in each situation where data will be collected. It is crucial that protected area planning teams be provided with gender training prior to information gathering, as sensitivity to women's issues and cultural situations will have an impact on the data.

The United Nations Research Institute for Social Development is launching a program to develop more data on how local communities can benefit from conservation measures in "buffer zone" areas. Other research institutions in the south should be encouraged to develop similar programs. Participatory workshops on ecological appraisal methodologies would enable local populations to decide on their own future and invest more interest in their local resources.

Conclusion

Strategies for improving basic needs of communities in protected areas must include the introduction of environmentally sound labor-saving technologies that free up women's time for more productive activities. Strategies must also ensure women's access to advances in agricultural science and technology. Perhaps even more important, strategies must build on women's knowledge base in such areas as seed selection, pest management, soil maintenance, and animal production.

Here are four ways to expand partnerships between women and protected areas. First, partnerships can be developed between donor agencies, governments, and local park management to support some key activities: training programs, research, and strategies to ensure that women's needs for access to and control over resources are being met. Second, national mechanisms can be set up to monitor how gender concerns are being addressed in local park management. Third, funds can be allocated to support research, publications, and wider communication on women and protected areas. Fourth, the steering committee convened by the IUCN to prepare for the Caracas Congress could be expanded to include those UN agencies and NGOs that specialize in environment and development. They could become an effective support group for national efforts.

Today, the world recognizes that basic human and social dimensions are central to the survival of our planet. The growing numbers of the world's women therefore have as valid a claim on our conscience and our resolve as our polluted air, contaminated seas, disappearing forests, and the deteriorating land base. It is up to us, women and men of conscience, to provide the vision and the leadership by valuing all our resources, including women.

Chapter 19

Redefining National Security: The Military and Protected Areas

Eustace D'Souza

Editor's Introduction:
In most parts of the world the defense services are a dominant force politically, socially, and economically. While their primary task is to defend the nation's political viability, the defense services are increasingly coming to recognize that political, economic, and ecological viability are closely interrelated. Yet they have seldom been systematically approached to provide their support for positive action in conservation of biological resources. Eustace D'Souza (Major General, Indian Army Retired) is a leader of the powerful and influential organization Generals for Peace, which promotes armed forces involvement in nonviolent action, including in environmental conservation. His chapter makes it clear that the military can make very significant contributions to national conservation objectives generally, and to protected areas more specifically. The military is ideally suited for this important and productive function. It has the leadership, discipline, training, organizational structure, administrative skills, mobility and communications. Military forces often are deployed in ecologically rich and often threatened areas, where their very presence suits them for conservation without having to relocate them specifically for this purpose. Through productive partnerships with the military, protected area agencies can ensure that significant areas of land important for conservation are managed in ways that are consistent with the national protected area system plan.

Introduction

In almost all parts of the world today, the military is a recognizable force politically, socially and, to an extent, economically. It is now widely accepted that in a holistic approach to security, military security is only one factor, and that humanitarian and environmental insecurity are likely to bring about intra- and interstate conflict.

The military has a unique nonviolent and productive role to play in protecting the environment, creating security and patterns founded on cooperation and not on confrontation. The current political climate clearly points to redefinition of national security to include the use of armed forces to protect and regenerate the environment. Since sustainable use and protection of natural resources are essential to ensure a reasonable quality of life, the role of the military in contributing to conservation and sustainable development as a means to attain national security needs to be considered seriously at this favorable period of declining international tension.

This study, based on the observations and experiences of 35 years of military service and 16 years of continuing contact thereafter, will aim to suggest how, in the current scenario, swords can be turned into plowshares, without blunting the cutting edge of the sword—at no stage can any nation afford to ignore the importance of the integrity of its borders.

The military can be deployed profitably for conservation tasks in a wide spectrum of ecosystems and for protection, regeneration, scientific research, monitoring underwater degradation, measuring harmful radiation levels, and managing defense lands. The navy for instance can monitor pollution of coastal and marine waters, poaching of maritime resources, and management of shore-based installations. The airforce is in a position to report on the status of inaccessible areas, forest fires, sighting of endangered species, and management of the large number of restricted areas like airfields. Army personnel can be profitably employed as wildlife wardens, and their bases can be managed to achieve conservation objectives. And, most important, military areas and installations can be eloquent exemplars to the nation, especially in the developing world.

The Military and Protected Areas: Success Stories from India

The military deployed in remote areas has the capacity to contribute to conservation in such diverse areas as water management, reforestation, afforestation, veterinary services, security, protective works, alternative energy sources, and scientific research. The success stories from India mentioned in the subsequent paragraphs at the micro, macro, and mega levels, all seen by the author personally, demonstrate that the military can be used with advantage for the protection of the environment.

Micro Examples
About 500 km south of Bombay lies the town of Kolhapur, a fast-growing industrial complex. On the southern outskirts is located a military unit with

a small permanent staff of 76 all ranks, joined by 200 personnel for two months per year for training. The hillocks surrounding the installation are bare, making training during the hot and dry summer months a problem. The commanding officer decided to green the degraded areas of the small can-tonment on a no-cost basis, with donated inputs. Within a period of four monsoons, 350,000 saplings were planted of trees endemic to the area, donated by an enthusiastic forest department. Thanks to regular care and security, the survival rate was 90 percent. Initial resistance from the men was easily overcome when the dedicated conservationist officers led from the front. Much-needed water was ensured by the construction of small but effec-tive check dams. A donated pump and pipes enabled the water to be pumped to a reservoir excavated on the highest hillock from where it was gravity fed to the saplings planted on contour bends. The mayor was so impressed that he donated a bore well to supply water to the nursery, overseen by a botanist provided by the forest department. Among the donations in kind received were a tractor with diesel fuel, liquid fertilizer, and security fencing.

This small cantonment, once devoid of trees, is now greened effectively to provide shade, aesthetics, and a habitat for wildlife. During his most recent visit the author discovered that raptors like the shikra had returned. Plans are afoot to introduce ungulates. Thanks to the efforts of these officers, who are here today and gone tomorrow, a new habitat has been created. Cattle are not permitted to graze but the tall grasses that have reappeared are cut and sold for a low price to cattle grazers as fodder and the funds so earned are plowed back into the project. The unit was declared the best in India, the com-manding officer was the deserved recipient of the Chief of the Army Staff's Commendation Card, and the local District Commissioner presented a silver trophy to the unit.

Not to be outdone, a regular army major of another local unit has jumped into the act. Involving cadets and school children, he has adopted another bare hillock, planted trees, and is constructing a 2-km nature trail. To ensure availability of water he has had three check dams excavated with bore wells, which will also attract winter migrants, and has planted endemic trees and constructed sit-outs along the trail. The local forest department and a pro-fessor of botany from the university nearby provide technical inputs. During a recent visit, the author saw a wildboar piglet in the tall grasses that now abound. This excellent project has received considerable media coverage.

Macro Examples

Kotah is a fast-growing industrial town and rail junction situated on the southeastern fringes of the Thar Desert, with the River Chambal flowing through it. It was once the Brigade of Guards Training Center and is now the home of an infantry division. Some conservation had taken place earlier prior to the arrival of this division led by a major general who, apart from being a mountaineer, was also an environmentalist. He set himself to restore tree cover to 500 ha of degraded land allotted to the formation as a training area on the outskirts of the town. He called it the Abhera Arboriculture Project. Through this project he would ensure shade for his troops during the searing hot summers; he would prevent soil erosion; and he would provide a habitat for wildlife that had vanished from there. In the first year 10,000 trees endemic to the area were planted, followed by an equal number in the subsequent two years with a good survival rate. A local NGO, noting these activities, sought permission to establish a gene reserve of endemic plants and the most fertile part of one corner of the project was allotted to them. Wildlife is beginning to return to this recreated habitat.

A similar effort has been launched in the degraded Chambal ravines on the outskirts of the cantonment. The troops provide the security, inputs, and construction of small check dams and the forest department the technical know-how and saplings. The area so regenerated has been used sensibly to construct two nature trails as part of the nature awareness program. Four nature trails have been created within the cantonment including two in the erstwhile ruler's reserve forest around the palace. Each trail has been allotted to a unit of the division and is of 2 to 3 km in length. On the Sambar Nature Trail, for example, can be seen sambar deer, wild pigs, primates, porcupines, peacocks, partridges, and many other species. Name boards in Hindi and English indicate the names of the flora and fauna seen. The trails on the banks of the Chambal start on the upper banks, descend down to the river's edge, and ascend again to shaded sit-outs overlooking the placid waters of the Chambal, where one is able to observe a wide variety of bird life. All these four trails are easily accessible, and this "natural" method of creating awareness is becoming increasingly popular. The general has also made excellent use of voluntary nature orientation camps within the area of his responsibility, helping to teach troops about conservation. The locations selected are the Tall Chappar Black Buck Reserve, the Ranathambore Project Tiger Sanctuary, and the famous Keoldeo Ghana Bird Sanctuary at Bharatpur. With the assistance of the State Forest Department and Bombay Natural History Society (BNHS), the subjects covered by the training course include do's and don'ts in sanctuaries, environmental terminology, the Indian Wildlife Protection Act, animal behavior, bird watching, water management, the importance of trees, pollution, and the basics of wildlife photography. These camps are open to all ranks and their families and have proved to be very popular. In the cantonment, the Kingfisher Nature Club has been formed, and a room in the officer's colony has been allotted as a clubhouse.

In its ongoing efforts to protect India's wildlife, the Army has set up three bird sanctuaries in Jammu Province, one on the Beas River and the other at the Hussainiwala Headworks. Wildlife sanctuaries or refuges are being developed in Bengdhubi (North Bengal) and the ecologically famous Loktak Lake in Imphal.

Mega Level

In North India is located an army corps in an area that, some years ago, was devoid of any significant vegetation cover. A succession of enlightened corps commanders, using military resources, succeeded in converting this desolate area into a veritable forest, providing much-needed shade, a habitat for wildlife, and aesthetic values. The large camps located there have become safe havens for wildlife—ungulates, wild pigs, primates, quail, partridge, and so on—because of stringent antipoaching orders and security.

At the army command level, the Northern Army Commander, aghast at the total deforestation of a mountain behind his headquarters, has reforested the whole area with troop labor and has taken the precaution of constructing an approach road to the highest point to locate a company of troops there sensibly for security but in actual fact to protect the new forest. The same army commander, since retired, persuaded the Kashmir State Forest Department to appoint army officers as honorary wildlife wardens.

Recently there has been a visible increase in such activities at army command level throughout India. Northern Command organized a week-long workshop in the Dachigam Wildlife Sanctuary, home of the endangered Kashmir Stag (Hangul) and the Himalayan black and brown bears, with the assistance of WWF India, BNHS, and the State Forest Department. Airforce officers and wives attended as well. Among the decisions taken were the appointment of designated army officers as honorary wildlife wardens; issue of kits to troops to enable local species of flora and fauna to be identified; formation of an environmental liaison unit to be established in the army command headquarters to coordinate all such activities and to liaise with other organizations for technical inputs; encouragement to officers wives to devote time to creating conservation awareness among troops' families; and a joint study to provide technical advice for this ecologically rich state.

Western Army Command has conducted a 10-day workshop and nature orientation camp for its officers in the Keoldeo Bird Sanctuary and in the Sariska Project Tiger Reserve to educate its personnel in conservation problems, protection of endangered species, and the importance of trees. Eastern Army Command, assisted by WWF India's West Bengal State Committee, organized a well-attended nature orientation camp in the Buxar Tiger Reserve where the problem of elephant migration was studied in view of the diminishing habitat for these pachyderms.

In the ecologically rich district of Kutch in the State of Gujarat, home of the Indian Wild Ass and the Great Indian Bustard, both endangered, the local army formation conducted a three-day seminar on the ecology of that area. The State Forest Department, the Ministry for Environment, BNHS, WWF India, and other NGOs participated. A wide range of subjects was covered. The main thrust was the saving of one of the largest grasslands, the "Banni," from degradation due to salinity and encroachment. Ways and means of assisting scientific research were covered. The protection from tourists of the flamingo colony in the Rann of Kutch, where thousands of Greater Flamingos migrate to breed after the rains, was also discussed. The formation has planted more than 10,000 endemic plants in the degraded areas of the cantonment in typical army fashion. This will be followed by similar projects to green the cantonment.

The structure of the military and the nature of its deployment do not value awareness—creating an insurmountable problem, given the interest shown by three successive prime ministers of India, each of whom ensured, in varying degrees, the need to protect nature. This culture trickled down the line to the armed forces, and orders emanating from the topmost levels of military leadership reached right down the line. It was therefore decided to direct the thrust at two important levels, the officer cadet and the middle-level officer, the future decision makers.

At India's National Defense Academy, officer cadets for all three services train together for three years in an ecologically interesting environment. All cadets are put through a three-day crash course in environmental ethics during midterm breaks. Participation in nature orientation camps conducted in selected biomes are encouraged and fully subscribed. NGOs conduct four-day camps for 300 cadets each, twice a year. The campus provides a made-to-order example of how the military can convert the once-bare hills into a forest. Experts from various conservation-oriented organizations are invited to make presentations and to interact with the faculty, staff, and cadets.

The Defense Services Staff College is located in the ecologically rich Nilgiri Hills (Blue Mountains) on the Western Ghats. Selected middle-level officers from all three services spend one year preparing to be decision makers. Some 400 officers on a selective basis attend this course in an area where clear-felling of endemic trees for potato cultivation has been going on for years, and where the introduction of the Australian blue gum, and wattle for the tanning industry, has created ecological problems. This institution of military learning provides a ready-made captive audience. Conservation is included in the syllabus, covering India's environmental problems. There have been a number of useful fallouts. A proposal to plant eucalyptus on the campus was stopped in time and endemic species planted instead. Students on arrival are encouraged to plant two fruit trees in front of their quarters and are required to care for them during their year's stay. A nature club for children has been established. During midterm breaks, an increasingly large number of officers are spending time in nearby wildlife sanctuaries. Prospective commanders and staff officers are advised to implement the arboriculture plan before construction of a new location commences.

All military centers and schools/colleges of military training are fertile areas for creating awareness. Such programs have and are being conducted by BNHS and WWF India, for instance at the College of Combat, Infantry School, Military College of Telecom Engineering, School of Artillery, Armored and Mechanized Infantry Centers and Schools, and Infantry Regimental Centers. This has resulted in increasing greening activities and a better awareness of the need to protect the environment.

Every year nearly 50,000 soldiers and airmen retire from the Indian Services. These form a valuable pool of trained, motivated, and disciplined manpower. In developing countries where labor availability makes environmental rehabilitation a problem, what better method than using this manpower for the protection of nature? India is perhaps the only country to tap this vast reservoir of retired servicemen for ecological work. Today there are three "eco-battalions" operating in the country: two in the degraded Himalaya and one in the Great Indian Desert. Each is commanded by a regular army colonel assisted by a small core of regular army personnel. The men are mostly ex-servicemen recruited from the area where they are required to operate, thereby ensuring a vested interest in their eco-duties. These units are financed by the Ministry of the Environment, though for recruitment, equipment, and training they come under the Ministry of Defense. These pilot units have done commendable work in afforestation, reforestation, soil erosion, and similar productive tasks.

The Military and Protected Areas: Examples from around the World

Although the author by virtue of his long-continuing close association with the Indian military has highlighted the efforts of the defense forces of India, there are a number of examples of the involvement of the military in other countries in the field of protected areas (see Table 19-1). United States Senator Sam Nunn is on record as suggesting that the possibility of the U.S. armed forces being mobilized to fight environmental foes like global warming and deforestation is very much in the cards. The senator is of the considered opinion that the USA's military mission should be redefined to include

Table 19-1. Examples of Cooperation Between Protected Areas and the Military

CUBA	Military representative on the national commission for environmental protection; creation of national parks/reserves in military areas; assistance in reforestation in difficult/mountainous areas; environmental education.
DENMARK	Conservation of wildlife/flora in army-controlled areas.
EGYPT	Protection of marine resources: coordination of environmental protection measures with civil authorities; expansion of green areas on defense lands.
FRANCE	Protection of forests; publication of a periodical on the army's involvement in environmental protection activities.
ITALY	Assistance in fauna protection; protection of flora and fauna against fire hazards.
JORDAN	Formation of special units of conscripts to assist the army in conservation programs.
NEW ZEALAND	Employment of the military on public-related services; provision of logistic support to the Department of Conservation; control of fishing.
NEPAL	Provision of guard force for Royal Chitwan National Park.
TANZANIA	Afforestation/reforestation measures in army areas; antipoaching measures; environmental education programs for youth conscripts.
THAILAND	Army involved in re-greening programs.
UNITED KINGDOM	Conservation management programs for defense lands; appointment of an environmental officer in the Ministry of Defense; involvement of over 5000 civil/military volunteers in joint conservation groups; protection of flora and fauna in overseas military establishments; publication of *Sanctuary*, a periodical devoted to military environmental activities in the UK and abroad.
UNITED STATES OF AMERICA	Management of military reservations for conservation; transformation of surplus military areas into protected areas; control of exotic species on Pacific Islands controlled by military; research on conservation-related topics; adaptation of technology originally developed for military uses to conservation uses.
VENEZUELA	Control of hunting and fishing; antideforestation measures; protection of marine and bird life by the navy; establishment of a scientific base on Isla de Aves; inclusion of environmental responsibilities at military training institutions.

research on environmental problems which, in his words, "pose a new threat to the world."

In October 1990, a draft resolution was submitted to the U.N. General Assembly by nine countries including India and Venezuela, calling on the Security Council to institute a study into how the potential use of resources hitherto available to the military could be used to support civilian endeavors to protect the environment.

It has been suggested to develop a series of case studies in which the military is having a positive influence on conservation of biological resources, as in Burma, Madagascar, Sri Lanka, China, Venezuela, the USA, the UK, and Zimbabwe. All military leaders who have demonstrated a sensitivity to environmental issues might well be brought together with conservation professionals to recommend how the defense services can be approached most effectively to promote conservation interests.

Conclusion

Based on the experience to date on mobilizing the military in support of protected areas, it is clear that there is a strong case to redefine national security to include the protective role of the military among other security considerations.

Why this is so may be gauged from various factors.

- The military, the world over, controls large areas of land for training, for border buffer zones, and for military installations. Very often such areas are of considerable biological and ecological importance, providing a safe habitat to a variety of flora and fauna.

- In some developing countries the military is actively involved in rural development programs.

- It is possible to incorporate conservation, ecology, and the environment into military training programs for a very "captive" audience.

- The excellent information acquisition services available to the military on land, under water, and in the air in a variety of fields could be applied to solving conservation problems, such as planning of protected area systems.

- In many developing countries, a large number of service men and women come from rural backgrounds, making them well predisposed for conservation.

- The deployment of the military in remote wilderness areas makes them suitable as *in situ* and *ex situ* contributors to conservation of bio-logical diversity.

* As the military are themselves dependent on a healthy environment for their wherewithal, they can be reasonably expected to have a seri-ous interest in resource management issues.

The structure, organization, training, leadership, and infrastructure of the various branches of the military make them ideal partners with their civilian counterparts in protecting the environment. The rigid mindsets of today that the military generally speaking are destroyers rather than protectors of the environment can and must be changed to allow for an acceptance of the new thinking.

In redefining national security it must be borne in mind that the military's contribution to protected areas must be in relation to their primary goal of ensuring the integrity of national borders and internal security. The Indian Armed Forces have been involved in hot and cold wars, internal and exter-nal security, counterinsurgency and antiterrorist operations, disaster relief, peacekeeping operations for the UN, and so on, without a break since 1948. Yet, they also have been increasingly involved in protecting India's fragile environment in a positive manner without blunting the cutting edge of the sword. This has been possible through the personnel interest and encourage-ment of its military leaders at all levels, its deployment in ecologically rich and remote areas, its infrastructure and, in some cases, its "eco-battalions." That the two roles are not conflicting but complementary must not be lost sight of. Governments need to remember that there is no common rule of thumb to achieve these redefined goals; they depend on the culture and ethos of each nation. But some of the common steps through which this important goal can be achieved are (a) acting as exemplars in locations normally occu-pied or controlled by the military; (b) advantageous use of the organizational structure, leadership, training, equipment, communications, and mobility inherent in this force; (c) utilization of the large number of training institu-tions to spread the message of conservation; (d) transborder military cooper-ation on protected areas issues facing both sides; (e) projects involving afforestation, reforestation, water management, reduction in pollution levels, use of nonconventional sources of energy, and protection of endangered wildlife; (f) assistance to scientists and NGOs; and (g) antipoaching mea-sures.

The success stories mentioned and the contribution to environmental pro-tection by a steadily increasing number of countries is a clear signal that the time is ripe for nations to redefine national security to include the protection of the environment by the military.

Part III

PARTNERSHIPS WITH COMMUNITIES

Chapter 20

No Park Is an Island

Ervin H. Zube

Editor's Introduction:
Significant strides have been made in the past decade toward recognition of protected areas as part of regional landscapes, leading to new partnerships with institutions responsible for managing the surrounding lands. Ervin Zube describes four different kinds of relationships between protected areas and local populations, including local participation in a protected area, protected areas providing services and assistance to local communities, accommodating traditional land uses within the protected area, and local participation in tourism activities. It is clear that the range of possible interactions between protected areas and surrounding lands are multiple, and this chapter includes a set of general guidelines that can direct the protected area planning process in a way that will build the most productive possible relationship with the surrounding lands. This calls for managing protected areas at a regional scale.

Introduction

National parks and other protected areas have frequently been described as islands. In island biogeography theory, this term has been used to explain species–area relationships, isolation effects, and species turnover (Shafer, 1990). It is also an apt description of the way many protected areas have been managed, as islands isolated from surrounding regional landscapes. While the appropriateness of the island biogeographic conceptualization is still being debated in the biological realm, this description represents a serious management misconception in other respects.

Protected areas are connected to their surroundings through ecological, economic, and cultural relationships. These relationships and interactions can be either positive or negative or, perhaps, neutral. Figure 20-1 illustrates some of the primary biophysical interactions between parks and regional landscapes. Polluted air and water, as well as exotic species and fire, can enter

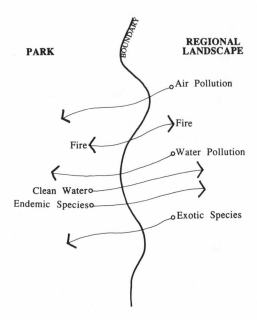

Figure 20-1. Biophysical Interactions

from outside the park. Conversely, clean water, fire, and endemic species can move from within the park to the surrounding landscape. Positive and negative interactions can occur on both sides of the boundary.

Another set of relationships requires attention, those involving human interactions with the protected area. Frequently, these interactions have been thought of in terms of tourism and poaching. While these are both significant, they do not address the issue fully. This chapter discusses efforts that are being made to enhance the quality of these interactions. In doing so I draw on two recent studies: the first, an international survey conducted to identify individual protected area initiatives; the second, a recent assessment of nontraditional units and affiliated areas of the United States National Park System (NPS).

Background

It is 30 years since the First World Conference on National Parks was held in Seattle; others were held in Yellowstone in 1972 and in Bali in 1982. A review of the proceedings from those conferences is informative about changing conceptualizations of the relationships of protected areas with their neighbors and with the regions of which they are a part. Stuart Udall's keynote address for the 1962 conference, "Nature Islands for the World," reflected a prevailing concept of parks as islands, isolated from the surrounding natural and cultural landscapes (Udall, 1962). At that time Udall was U.S. Secretary of the Interior, the senior federal administrator responsible for the NPS.

The Proceedings of the 1972 Second World Conference suggests the beginning of a broadening concept of park–regional landscape relationships. George Hartzog, then director of the USNPS, put the issue into perspective when he said: ". . . it is highly important that parks should not be treated as isolated reserves, but as integral parts of the complex economic, social, and ecological relationships of the region in which they exist" (Hartzog, 1972).

Among the set of approximately 35 recommendations approved by conference participants, only one addressed this issue. It recommended that compatible land use practices be implemented outside boundaries of national parks.

The 1982 conference in Bali marked a major advance in recognition of the ecological, cultural, and economic contexts within which national parks exist, and of the necessity to take these contextual issues into consideration in planning and management. In his summary of the conference, Kenton R. Miller, then chairman of the IUCN Commission on National Parks and Protected Areas, noted that, "protected area management can be characterized by its 'insular approach'; goals are selected and means implemented within boundaries of designated areas with little or no regard for surrounding lands and peoples. As a result, social and economic conflicts arise along the margins of reserves, and popular awareness and political support for protected area programs are diminished" (Miller, 1984). He noted that many case studies were presented at the conference describing how this problem was being addressed. Miller's comments reflected the spirit of the time. Only two years earlier, IUCN, together with UNEP and WWF, had published the *World Conservation Strategy* (IUCN, 1980), which called for linking conservation and development.

If protected areas are natural or ecological islands as has been suggested, there are now serious attempts to build natural and cultural bridges, causeways, and corridors to surrounding natural and cultural landscapes. In fact, protected areas are components of larger systems and, as illustrated in Figure 20-1, are influenced negatively and positively by factors from outside their boundaries. In turn, as noted by Miller in 1982, protected areas have external influences on the surrounding landscape. Miller noted the beneficial influences; but, from the perspectives of local populations, protected areas can also have harmful effects. Examples are endemic carnivores moving beyond boundaries or the loss to local people of opportunities for traditional uses of lands included within the protected areas.

Protected areas must be conceptualized and managed as parts of dynamic landscape systems. This systems approach extends beyond the concept of protected area systems—the categorization of areas identified as important to preserve the diversity of ecosystems within a nation and the world. The emphasis here is on the concept of individual protected areas as parts of larger natural and cultural landscape systems. This concept is receiving increased support, but it has not received the attention it demands as a critical component in protected area planning. This is understandable, because it is difficult to accomplish. It involves planning with a broad-based interdisciplinary team approach. It requires planning by education and persuasion, and by building alliances and partnerships that transcend park boundaries. Furthermore, there is no guarantee for success. Nevertheless, few, if any, viable alternatives are available.

Thinking Regionally

What are the relationships among protected landscapes and the broader natural and cultural landscapes of which they are a part? The international survey conducted in 1986–1987 explored ways in which protected areas were attempting to facilitate positive interactions with local populations (Zube and Busch, 1990). It provides one insight into the question of regional cultural relationships. The 1990–1991 study of recent additions to the U.S. National Park System provides another example of efforts to establish positive interactions with regional landscapes, both natural and cultural.

The International Survey
The purpose of the international survey was to identify approaches being pursued to build bridges between protected areas and local populations. A literature review and subsequent mail survey identified 99 areas actively working to foster positive relationships with local populations within their regions. This is at best a sampling of what is actually happening, because more information becomes available each year about similar initiatives in other areas. The 99 areas represented 39 countries ranging from Australia to Zimbabwe. Four general park–local population models were identified, each model representing activities of a similar nature (Figure 20-2).

- *Local participation in protected area*—ownership of protected area resources, living within the protected area, serving as protected area administrator or employee, and serving as members of advisory or policy review committees
- *Park providing services and assistance to local communities*—education

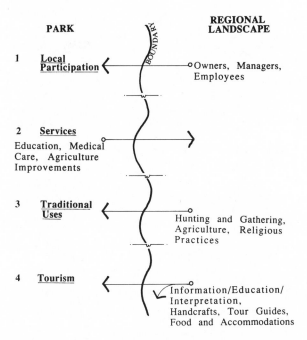

Figure 20-2. Park/Local Population Interactions

programs, technical assistance for agriculture, new economic activities, and healthcare

- *Accommodating traditional land uses within the park*—hunting and gathering, agriculture, religious practices, and pastoralism
- *Local participation in tourism activities*—sales of handcrafts, providing in-park education programs, serving as tour guides, and providing support services for tourists such as transportation, housing accommodations, and food

When the four models of cultural interactions depicted in Figure 20-2 are added to the biophysical interactions depicted in Figure 20-1, the complexity and diversity of park–region interactions become evident, even when depicted in the simplified abstraction of Figure 20-3.

New Protected Area Concepts in the United States
The second study involved 14 protected areas in the United States. These areas varied considerably in location, resources, and resource ownership and management. Some were established to preserve important historic structures

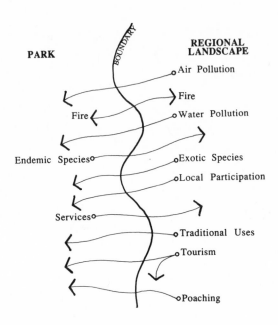

Figure 20-3. Biophysical and Cultural Interactions

and sites, others to preserve important natural areas, and still others to preserve a combination of historic and natural resources. Some were near urban centers and others were remote, but all are involved in one or more formal and informal partnership arrangements with nonprofit organizations, private businesses, citizen groups, religious groups, and other units of government (including other federal agencies as well as state, city, and town governments). In contrast with traditional national parks where the federal government owns and manages all park lands and resources, these areas accommodate shared ownership and management of resources. The purpose of the study was to identify the planning and management practices and procedures that contributed to or hindered the success of the protected areas. It is difficult to conceive of any one of them as an island.

The following brief descriptions of several protected areas illustrate ways in which they differ from traditional NPS units and the kinds of partnerships that have been established to address local concerns and to relate the area to the regional natural and cultural landscape.

Pinelands National Reserve is an area of approximately 405,000 ha of forest and wetlands in the urbanized northeastern corner of the United States. The NPS provided technical assistance for development of the management plan and financial grants for purchase of lands and waters within the reserve. Planning and management of the reserve is the responsibility of an appointed 15-member commission consisting of seven members representing political units within the region, seven members from the state of New Jersey representing environmental interests, and one representative from the NPS. There is active and significant regional involvement in preserve planning and management.

El Malpais National Monument and National Conservation Area in Arizona is a protected area containing a mosaic of geological and cultural resources including volcanic cones, lava tubes, ice caves, unusual biological resources, and many archaeological sites. It encompasses an area of approximately 152,000 ha divided between the NPS and the Bureau of Land Management (BLM). The NPS mission is resource preservation and recreation. The BLM mission is multiple use of resources. The spatial relationship of the lands owned and managed by the two agencies is similar to the biosphere reserve concept, with a preserved NPS core area surrounded by a limited-use BLM area. There are, however, other partners—the Acoma Indian Reservation on the east boundary and the Ramah Navajo Indian Reservation on the west. Within the area are places sacred to the Acoma and Navajo peoples. Congress specified that boundaries be sensitive to the needs of the Indians, that access be maintained to sacred and traditional-use sites, and that they be protected from recreational users of the area.

Other examples include the Ice Age National Scientific Reserve and the 1500 km long Ice Age National Scenic Trail. The reserve involves NPS partnerships with a state natural resources agency, which owns and manages the National Reserve, and with a private nonprofit foundation that acquires and manages land for the National Scenic Trail. The NPS provides technical assistance and annual contributions to the costs of operating both areas.

Another protected area, San Antonio Missions National Historic Park, involves a cooperative agreement between the NPS and the Catholic Archdiocese of the city of San Antonio. The primary historic resources of the park are four active 18th century mission churches. The Archdiocese maintains responsibility for the interiors of the churches and the NPS has responsibility for the exterior of the buildings and surrounding landscape.

At Santa Monica Mountains National Recreation Area, a developing relationship between the NPS and local real estate developers is focused on protecting wildlife corridors that traverse both private property and the National Recreation Area. The intent is to convince developers to donate these important corridors on private land to the park in exchange for park support in design, planning, and gaining community approval of environmentally sensitive development projects on the boundaries of the park.

Conclusion

Analysis of those elements that contributed to success of protected areas included in both the international study and the study of new units and affiliated areas of the USNPS provides a basis for a set of guidelines for establishing positive interactions between protected areas and surrounding cultural landscapes. Each protected area has its own unique ecological, cultural and economic relationships with the surrounding region, and these must be identified and considered in an assessment of the potential of each guideline to contribute to successful planning and management strategies. These guidelines are intended to assist the planning process in a way that will reveal and be sensitive to those relationships.

- Identify all critical interactions (physical, biological, and cultural) that link the protected area to local populations and regional landscapes.

- Search for and gain understanding of the meanings and values that residents within the region attribute to the protected area and the region.

- Inform local populations about the national and international significance of protected area resources and strive to develop a sense of pride in the protected area and the region.

- Search for and implement ways to provide benefits that compensate for costs experienced by local populations because of the protected area.

- Conduct planning as an open process that provides opportunities for all interested parties to express their opinions and views about the future of the protected area and the region.

- Avoid preconceived ideas about how things have to be done. Few problems have only one solution. Consider local solutions for local problems.

- Appoint area managers with skills required for the specific protected area. Some areas may require greater skills in working with government agencies and local interest groups than others.

- Use advisory councils with members who can contribute to maintaining open communications with local populations and sensitivity to local values within the region.

In conclusion, protected areas are parts of larger ecological, cultural, and economic systems; they are inextricably linked to the regional landscapes of which they are a part, and the elements of those linkages are dynamic, not

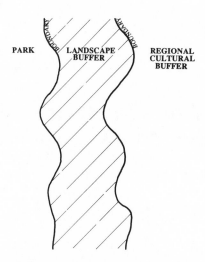

Figure 20-4. Biocultural Landscape Buffer

static. This requires an approach to protected area management and planning that is regional in scope. Perhaps the buffer concept should be expanded to include a cultural landscape adjacent to protected areas, a buffer devoted not only to protecting significant park resources but also one that demonstrates economically viable, ecologically compatible, and culturally acceptable land use activities (Figure 20-4)—a buffer that addresses the needs of both the protected area and adjacent local populations.

References

Hartzog, George. 1972. Management considerations for optimum development and protection of national park resources. 155–161. In H. Elliott (ed.) *Second World Conference on National Parks* Morges, Switzerland: IUCN.

IUCN, 1980. *World Conservation Strategy*. Gland, Switzerland.

Miller, Kenton R. 1984. The Bali Action Plan: a framework for the future of protected areas. 756–764. In J.A. McNeely and K.R. Miller (eds.) *National Parks, Conservation, and Development*. Washington, D.C.: Smithsonian Institution Press.

Shafer, C.L. 1990. *Nature Reserves*. Washington, D.C.: Smithsonian Institution Press. 189 pp.

Udall, S.L. 1962. Nature islands for the world. 1–10. In A.B. Adams (ed.) *First World Conference on National Parks*. USDI, NPS, Washington, D.C.

Zube, Ervin H., and Miriam Busch. 1990. Park–people relationships: an international review. *Landscape and Urban Planning*. 19:117–131.

Chapter 21

Protected Areas and the Private Sector: Building NGO Relationships

Ruth Norris and Laura Camposbasso

Editor's Introduction:
The role of NGOs in supporting protected areas has been growing rapidly over the past decade. Effective relationships between protected areas and NGOs can bring significant additional resources to the task of managing protected areas. Building organizational capacity to make these relationships work is key to the contribution that NGOs might be able to make to such a partnership. Ruth Norris, a consultant affiliated with The Nature Conservancy and other organizations, and Laura Camposbasso of WWF-US, bring many years of experience in building NGOs in both developed and developing countries. They describe how effective partnerships can be established, both with agencies responsible for managing protected areas and with international sources of financial support. This chapter provides wise advice on how NGOs can develop the most appropriate partnership relationships for their particular needs. Its messages are also relevant for governments and industry.

Introduction: Partnership and Organizational Development

The past two decades have brought an explosion in the number of nongovernmental organizations (NGOs) in developing countries dedicated to protected areas and conservation. For donors and developed-country organizations this has meant greater opportunities for implementing conservation projects. The NGO revolution also has brought challenges. Relationships between developed-country organizations and developing-country organizations have had their ups and downs, as have relationships between protected area agencies and the private sector.

At their best, interorganizational relationships and collaboration generate increased resources, public support, and flexibility for protected area management. At their worst, disagreements and competition lead to the formation of more NGOs to "set things right." The ability to develop productive relationships is a key attribute for any conservation organization. It ranks with clarity of mission, ability to develop focused projects, and ability to attract necessary human and financial resources as crucial elements of a successful organization. As practitioners of organizational development, we work with NGO and agency leaders to support them in crafting these elements.

Many of the organizations with which we work are in transition, having been founded with a specific project in mind, and struggling—almost as an afterthought—to create the structure, skills, and relationships necessary to make a long-term contribution to conservation. We believe that effective leadership in this context requires a conscious effort to balance power and negotiating skills. A collaborative relationship can work in the long term only when its terms of reference are clearly defined and agreed on; and when both organizations invest substantial effort in managing the process of the relationship as well as its products.

Symmetrical Relationships and the Balance of Power

The construction of interorganizational relationships requires time and a symmetry of intent and attitude. If one organization is funding another, developing a balanced relationship is especially difficult. "How can you have a mature relationship when you depend on them for money?" asked one of our respondents.

The first step toward a balanced relationship is to recognize that the dependence is mutual. Both partners in the relationship have needs. Most developing-country NGOs working in protected areas have needs for on-the-ground collaboration with protected area agencies on the one hand, and funding and technical assistance from their northern partners on the other. Northern NGOs seek relationships to develop and implement projects, and to gain information and insights about environmental situations in countries other than their own.

As the organizations begin to work together, two characteristics have more effect than any others: the planning capability of each organization, and each organizations's attitude toward the relationship. A balanced relationship is likely to develop when both organizations have good planning capability or when the relationship supports creation of such capability. (We use the term to include both development and implementation of strategically selected activities.) Unless each organization can define its mission and objectives, demonstrate its qualification in the selected areas, and negotiate partnerships that support rather than derail, activities selected by one party to the relationship are likely to dominate the agenda without necessarily serving both organizations' purposes.

Strong organizations have recognizable characteristics.

- Well-defined mission and strategy
- Committed and active board, management, and staff
- Programs and projects that embody the mission
- Ability to secure human, financial, and physical resources
- Identified constituency
- Supportive organizational structure and operational systems
- Effective management processes
- Systematic monitoring and evaluation

This seems obvious, but in our work we have experienced so many instances of program derailment caused by the "funding available" syndrome that we will consider the issue in some detail. Besides planning capability, attitudes and policies of the participating organizations have a profound effect on the symmetry or lack of same in the relationship. The organization providing the funding may restrict the type of activities recipients may undertake, or require strategic planning reports that recipients are not prepared to write. This again is a fundamental issue with a fairly simple response but an obvious need, judging from experience, for a closer look. Clearly, when entering into a collaborative or financial relationship, both organizations should go through a questioning process about their long-term ability to work together. We provide here a framework for that analysis, and we will come back to the question of making a relationship balance over time when the initial analysis indicates asymmetry.

We have found significant agreement on what makes a relationship successful or unsuccessful. Most NGOs associated success with mutual understanding and respect for each other and the joint project, honesty about concerns and needs, and a balanced sense of each other's role. A good relationship, they said, requires a time commitment from both parties, as well as discussions and mutual agreement on purpose and on issues such as which organization was to take credit for the success or failure of projects.

In effect, we have been asking how the interest and priorities of scientists, local residents, protected area managers, agency officials, conservation organizations, and the diverse community that support them can be translated into workable agreements and relationships that achieve both cordiality and results. We have looked at NGO(donor)–NGO(implementer), donor agency–NGO, and NGO–protected area agency efforts that can span a scale between two distinct relationship types: interventionist and interactive. We find a strong preference for the more symmetrical interactive mode. However, we have also found a significant number of organizations that have learned techniques for working with interventionist collaborators in a way that produces both results and mutual satisfaction.

Characteristics of Interventionists and Interactive Relationships

A relationship runs on the "currencies" of money, information, status, credibility, time, power, and on-the-ground conservation results. These currencies may be freely given, contracted for, even extorted or demanded by either partner. Cultural values and perceptions both of the currencies and the manner of exchange often get us into trouble, most often when we are least aware.

Organizations also have distinct "corporate cultures," values, and perceptions that drive their behavior and attitudes. The individuals who represent organizations bring their own personal leadership styles, values, and ideas about the various "currencies" to each interaction. Sorting out these styles, and becoming aware of the differing ways we see and use the various currencies, helps us predict and manage points of contention.

It probably is possible for an organization to be just plain wrong. But it is usually not helpful to tell them (us) so. We hope that by understanding differing relationship styles, organizations can become more effective at offering feedback and negotiating for needed support. All styles have positive and negative attributes. An organization that is committed to working in a certain way does well to seek out partners with similar preferences, or at least be aware, when entering a relationship with an organization of differing style, that conflict is likely. Planning for it will help keep the heat down.

The Interventionist Model

The interventionist relationship is characterized by dominance and control (Katz, 1956). The organization with the stronger agenda uses the powerful carrots of information, funding, technical and organizational assistance, and whatever prestige it is able to confer, as tools to recruit other organizations to carry out the agenda.

The interventionist model has been labeled "shallow participative style." "In this case, environmentalists, as holders of a true knowledge on ecological issues, eagerly approach local communities, who are ignorant of these situations. This style has the risk of . . . selling the environment as other companies sell kitchens, cars, or soap" (Evia and Gudynas, 1989).

When power is concentrated in one organization, the relationship is often rigidly defined. The dominant organization has experience in dealing with certain situations and has generally found specific types of interventions to be effective. This in turn causes the dominant organization to tend to categorize any conflict or problem as one of the situations it knows and can deal with, and leaves little flexibility for recognizing, let alone resolving, unexpected problem situations.

Problems frequently associated with the interventionist mode include boom and bust funding cycles, with availability at times in excess of the recipient organization's absorptive and performance capacity, at other times insufficient to even keep its doors open between projects. The "more reports than projects" syndrome is common when the dominant organization has reporting and accounting systems and needs beyond the recipient's capability to understand, let alone work with.

Why would a developing-country organization or a protected area agency enter into the relationship just described? It does have its advantages. The receiving organization has a secure source of funding and technical assistance so long as it conforms with the objectives of the provider. The risks associated with program selection, and the responsibility in case of failure, fall much more heavily on the shoulders of the provider. The receiving organization, should it find another sponsor or decide to develop capabilities of its own, can walk away at any time. It can use this power, by mutual acquiescence or one-sided threat, to gain concessions. The interventionist relationship usually is less ambiguous in its terms than the interactive.

The Interactive Model

The interactive relationship is defined by a two-way flow of the various currencies. Observation, analysis, and communication allow the distinct organizations to be clear with themselves and each other about their individual goals and needs, and to define a mutual working relationship within those areas where those goals and values are shared. Each party accepts that the other's goals and values are legitimately different from its own. The parties are interdependent. They have a clear and mutually agreed decision-making process.

Most NGOs share our preference for interactive relationships, but agree that they are more difficult to develop and maintain than interventionist relationships, with more opportunities for communications breakdowns and lack of results.

Between these two extremes are numerous variations. Indeed, it is possible for any organization to be exceedingly dominant in some aspects of its programs (for example, project objectives or accounting requirements) and exceedingly flexible in others (such as field procedures). However, very few organizations devote significant time to analyzing, let along making clear to other organizations with whom they work, which of their operating principles and practices are unavoidably inflexible, and which may have room for negotiation.

Developing Effective Interactive Relationships

General Comments and Recommendations

Our first recommendation to organizations wishing to build capacity for interactive relationships is to invest in self awareness. Organizations, like humans, have differences in perceptions, values, experiences, and expectations. In any given situation, we select and recognize factors consistent with our values and ignore factors that conflict. It may be more practical in the long run to recognize and adapt to our own and others' archetypal styles than to attempt to change them.

Relationships tend to work best in a "cards on the table" atmosphere. According to Peters (1991), "Sharing strategic information is arguably the clearest indicator of intent to open the partnership floodgates. . . . Until information becomes widely available, all other aspects of 'partnership' remain stuck at the stage of lip service or less." We have found enormous frustration on this issue from both donors and recipients.

Again quoting Peters, "Second only to sharing strategic information in making partnership a reality is involving (partners) in planning." We find here a good amount of rhetoric, but little consensus on what this would actually entail. A meeting now and again? Full membership for partner representatives on management teams? "The fact that the rhetoric of partnership outstrips the reality is really no surprise; not so long ago we didn't even have the rhetoric" (Peters, 1991). However, both northern and southern partners can expect to achieve higher levels of satisfaction if they will negotiate clear terms of strategic information sharing and involvement in each other's planning process.

We recognize that these are sensitive issues that have led to confrontations in the past. Participation in each other's planning process may not always be advisable. Northern organizations may find that unilaterally inviting partners to play a meaningful role in their internal strategic processes, without any expectation of reciprocity, improves the quality of their relationships.

The differing cultural perceptions of time and its value may be an issue. An organization may not be able to find time for telephone conversations and informal networking after a grant has been given, being by then stressed with review and processing of new grant applications. We have found a great deal of frustration and distress, even assertions of bad will or faith, generated by unmet communications expectations that appeared to result from one party's failure to "get to" intended actions rather than from intended withholding. Time management techniques, well known and widely available, should be more widely applied to resolve this problem.

We have found a widespread feeling that there is a better understanding now of the differing roles of various types of organizations and agencies—government, advocacy organizations, people's movements, project implementing organizations, policy research organizations, and so forth. However, the corresponding need for donor organizations to support a diverse environmental community is sometimes overshadowed by donors' inability to research and know a large number of organizations in every country where they work, and a tendency toward "one-stop shopping" that sometimes results in a single recipient organization being pressured to perform the functions of a whole social movement. Developing-country organizations are in a particularly advantageous position to address this problem with the formation of umbrella organizations.

Not surprisingly, most developing country NGOs feel that long-term relationships (more than five years) are preferable to short-term relationships. The relationship of trust that builds between particular staff members of the organizations helps both sides to invest seriously in personnel selection and retention.

We believe that organizations calling for change should begin with a willingness to make unilateral changes. One position frequently taken by our southern colleagues, that we believe it is time for northern organizations to accept unconditionally, is the obligation to use our expertise and clout to make changes at home. Energy and resource use in industrialized countries is far in excess of its proportion of the world's population. We as organizations with political voice must participate in such issues as energy conservation and the management of our own forests, as a condition of establishing our legitimacy in the global debate.

Recommendations Specific to Northern NGOs

"If we create too much of a dependence on us, paying per diem, buying vehicles, even paying salaries, we create an unsustainable situation. The relationship begins to be driven by dependency rather than by results."

The speaker, one of our colleagues in a U.S. NGO that provides financial and technical assistance to developing-country organizations working mostly in rural development, some in protected area buffer zones, cautioned us that this level of intervention often arises from day-to-day administrative convenience. His organization maintains an informal "checklist" of indicators that help predict the collaborating organization's ability to enter into an interactive relationship. These include rate of staff turnover, difference in the two organizations' political agendas, closeness or distance of the recipient organization with regard to the ultimate target population, whether the recipient organization's decision-making structure is field based or capital-city based, and whether the recipient organization shares key donor organization values and guiding principles such as gender equality, local participation, and sustainability.

The point is not to reject those that fail to "measure up" but to be informed of areas where conflict is likely to occur, or extra support is likely to be needed, and be prepared to manage it.

Donor organizations by definition are not driven by desperation for mere survival. We need to remember what panic feels like. When dependency is still an issue, it colors all communications. The panicked person is not likely to absorb much technical assistance. Panicked people have a single, overriding concern: building security. Organizational behaviors that build security, and leave more room for other messages to get through, include clear statements about the nature of the commitment and expectation, frank discussion of financial situations, and consistency. We often violate that last unintentionally, giving the impression early in a relationship that the collaborating organization will have ready access to us, our staff, our advice, our time—and leave them wondering whether they are about to be dropped when all that has really happened is that day-to-day pressures have overrun our good intentions.

What northern organizations can do to build capacity for strong, empowering relationships is not that hard to figure out. But it is hard to put into practice. Take time to do homework, understand cultural difference, make resources and expectations clear, and plan for long-term relationships. Work with umbrella organizations, limit the number of relationships, and take other steps to minimize the threat of handling more projects than can be understood and attended. Recognize that even the idea of generally beneficial management principles is a cultural value.

Recommendations Specific to Southern NGOs
Clearly we are not, and for that matter no one is, in a position to tell organizations what they "should" do. Every organization is the keeper of its own destiny and free to follow its own philosophy, agenda, and politics. Nonetheless, if an organization wishes to have a more open and productive relationship, certain kinds of steps are demonstrably productive toward that end.

The first step away from dependency is to focus on strategic planning. An organization can be strong only to the extent that it is sure of its mission and goals. To further them, it must develop negotiation skills. In our research, we have collected significant anecdotal evidence of relationships that began in confrontation but still worked out to mutual satisfaction when both organizations committed to a process of avoiding "denunciation" or "debate" and instead focused on negotiated means of meeting mutual needs. Of course, negotiation is possible only when the organization has invested in planning sufficient to organize its priorities and make its case for them.

Maintenance of contact is a two-way street. Be informed about the business of grant making and technical assistance, and the varying policies of different organizations. Often, misunderstandings occur that individuals within the organization see as personal or unfair, that are simply part of the business. The two most often cited are responsibilities for accounting for granted funds and evaluation processes. Ask questions up front. Offer information about your own policies and procedures, where you can be flexible and where you are constrained by laws or policies beyond your control, and ask collaborators to do the same. Talk, early in the relationship, about areas that are likely to cause conflict if left until they arise: ownership of and credit for success, policies for making press announcements, how and under what conditions the relationship is renewed or terminated.

Seek coalitions and umbrella organizations that will be in a more powerful position to negotiate than working alone. Diversify. Invest time in maintaining contact with a wide community of potential supporters even when you are not seeking immediate direct support.

Use your power to help collaborators choose successful strategies. Northern NGOs are not necessarily homogeneous or consistent—they work with a variety of southern NGOs and develop dynamics that may not perfectly fit any one. The same one may be supporting both paternalistic welfare and radical empowerment. Feedback from the field is crucial. They, like you, need to generate more projects, process more paper, and render ever more accounts, especially if they receive government funding (Elliott, 1987). This can be a source of mutual dialogue.

"Each person's view is a unique perspective on a larger reality. If I can 'look out' through your view and you through mine, we will each see something we might not have seen alone."

References

Agarwal, Anil. 1990. The North-South Perspective: Alienation or Interdependence? *Ambio* 19(2):94–96.

Dichter, Thomas W. Let's Make a Deal: The Third World Doesn't Want Charity. *Grassroots Development* 11(1):38–39.

Elliot, Charles. 1987. Some Aspects of Relations Between the North and South in the NGO Sector. *World Development* 15:57–68.

Evia, Graciela, and Eduardo Gudynas. 1989. Beyond Peoples Participation: Different Styles, Different Ethics—A Latin American Vision. Reprinted in *Information Notes-ANGOC* June, p. 6, excerpts from the paper in *Environment and Development*, CIPFE.

Garilao, Ernesto D. 1987. Indigenous NGOs as Strategic Institutions: Managing the Relationship with Government and Resource Agencies. *World Development* 15 (Supplement), 113–120.

Katz, Robert L. 1956. Human Relations Skills Can Be Sharpened. *Harvard Business Review* July–August, 82–93.

O'Brien, Jim. 1991. How Can Donors Best Support NGO Consortia. *Grassroots Development* 15(2):38–40.

Paul, Samuel. 1988. Governments and Grassroots Organizations: From Co-Existence to Collaboration. 61–71. In J.P. Lewis (ed.) *Strengthening the Poor: What Have We Learned?* Overseas Development Council, London.

Peters, Tom. 1991. The Boundaries of Business: Partners—The Rhetoric and Reality. *Harvard Business Review* September/October, 97–99.

Speth, James Gustave. 1990. Coming to Terms: Toward a North South Compact for the Environment. *Environment* 2(5):16–21; 41–43.

Chapter 22

Innovative Partners: The Value of Nongovernment Organizations in Establishing and Managing Protected Areas

Annette Lees

Editor's Introduction:
While the previous chapter describes experience primarily from the Western Hemisphere, Annette Lees, Associate Director of the Maruia Society, describes partnerships between NGOs and protected areas in the Pacific. Experience in this part of the world provides a slightly different perspective from that of the Western Hemisphere, because many of the island nations of the Pacific have low populations and small protected area institutions. In such settings, it is often up to local communities to decide what they need to conserve. Many NGOs in developing countries have especially strong links with local communities, being able to deliver effective, useful assistance to people in remote regions with a flexibility and speed that governments may be unable to match. A mix of development and environment NGOs working on the protected areas program of the Solomon Islands shows how a multitude of skills and resources can be mobilized to help design protected areas that meet the twin requirements of nature conservation and people's needs.

Introduction

In a world where most national governments regard themselves as suffering economic difficulties, care for protected areas, their selection, promotion, and management, competes for funds and attention with other national economic objectives. In these circumstances, it is the commitment of individuals to the protection of nature that keeps protected areas alive. Many of these people are found working outside governments in private organizations.

Nongovernment organizations have made significant contributions to protected areas. They commonly bring qualities of innovation, commitment, flexibility, and a history of community-led solution-finding to the complex task of successful management of protected areas. Their skills are an important complement to the role of government, a partnership that needs fostering through mutual respect and resource sharing. This chapter discusses a selection of approaches being taken by nongovernment organizations in the management of protected areas, highlighting the unique benefits (and problems) that NGO-initiated work presents.

Strength in Diversity

The growth of protected areas has been relatively slow in most countries, due, in good part, to the lack of involvement and support from key sectors of the public in protected area management. NGOs, representing a multitude of different interest groups, offer an opportunity to reach key sectors of the public to foster support for protected areas. At a community level, local NGOs can provide rapid access to the heart of a community's development concerns, providing an insight into potential opposition to a protected area proposal or future problems with reserve design and management. Nationally based NGOs can reach the wider public through well-researched publicity campaigns on the need to support a particular protected area. A sample of this diversity of NGO contributions is discussed below.

Environmental NGOs
In New Zealand, all of the major protected areas established in the 10 years since the World Parks Congress held in Bali in 1982 have been initiated through the public education and lobbying efforts of New Zealand's active conservation NGOs. Since 1981, two new and important national parks have been established, a World Heritage site declared, protected status given for large areas of wildlands including a number of key ecologically representative areas, mining evicted from national parks, and incentives established for protection of nature on private land. The conservation NGOs have led the public debate on these issues by well-researched campaigns promoting protected area establishment, and by coordinating and harnessing the growing public concern for the last remaining stands of representative forest types and natural landscapes in the country. Using the talents and commitment of professional staff, the NGO influence has put conservation firmly on the agenda of New Zealand's political parties. Globally, such NGO initiatives are not unusual.

Cultural Survival

NGO groups concerned with the rights of indigenous people have done much to publicize their plight in the face of large-scale development in the natural lands they inhabit. Protection of natural resources managed by indigenous people can be linked to wider protection of nature. Indigenous people's support groups have helped to highlight the importance and value of incorporating indigenous knowledge of ecosystems in the design and management of protected areas. Cultural Survival, an NGO concerned about indigenous people's rights, has also drawn attention to inequities of protected areas establishment where indigenous people are expelled from reserves or are allowed to remain only by retaining a subsistence lifestyle (for example, Gordon, 1990). Defending the basic rights of people who live in and around protected areas is an important contribution to the long-term stability of reserves. Alienated people may be hostile to a conservation program in which they have been unable to participate, so caring for and accommodating their needs at the onset of protected area development is essential. NGOs experienced in indigenous people's concerns can provide valuable assistance in reaching just solutions in these circumstances.

The Power of Partnership

The strength of many developing-country NGOs is in their strong links with communities. Many of these groups have evolved as community development organizations such as village water cooperatives, education and health groups, village women's associations, tree-planting groups, and so on. Born through community concerns, these organizations are frequently able to deliver effective, useful assistance to people of remote regions with a flexibility and speed that governments may be unable to match.

Until recently, most developing country community groups have focused on development, not environment. This was highlighted by a 1990 Commonwealth Foundation NGO Forum in Harare where NGOs from 48 countries met to discuss environmentally sustainable development. The Commonwealth Foundation aimed to invite one development-oriented NGO and one environment NGO from each country but was unable to find enough local groups to meet the second category. So, only 13 of the 170 delegates represented environment NGOs. This graphically illustrates how local energies in developing countries have had to be prioritized. The emphasis of these NGO groups is widening, however, as the significance of nature protection to rural people becomes increasingly apparent. This has stimulated interest in forming links with developed-country environment NGOs who have access to funds, technical expertise, and international lobbying power. Where the two groups can work together to assist people and governments to solve often interdependent environment and development problems their combined strengths can be extraordinarily effective.

An example can be seen in the Solomon Islands. Successfully establishing protected areas in this country is similar to most countries in that it is dependent far less on promoting the biological values of identified areas than it is on meeting the development needs and cultural aspirations of the local people. In the Solomons, indigenous people communally own and control more than 90 percent of the land and resources. Less than one percent of the land is formally protected. To achieve much needed forest protection the Solomon Islands Development Trust (SIDT), a local NGO that specializes in nonformal village education, has linked up with several international conservation NGOs, including the New Zealand-based environment NGO, the Maruia Society. With the International Council for Bird Preservation, the three organizations jointly conducted a national ecological survey for potential protected areas in the Solomon Islands (Lees, 1991). Follow-up work to the survey is proceeding, combining the village-based development and education skills of SIDT with the funding and information access provided by the Maruia Society and a fourth group, Conservation International.

The mix of development and environment NGOs working on the protected areas program brings a multitude of skills and resources to help design reserves that meet the twin requirements of nature conservation and people's needs.

The benefits of a partnership between development and environment NGOs extend beyond the enhanced effectiveness of a protected areas program. Sharing skills and experience also influences the objectives of environment groups, bringing environmental awareness to the heart of community development and helping environment groups understand the importance of caring for human needs. Placing people at the center of solutions to environmental problems is as critical in developed nations as it is in developing countries.

Liaison between Local Communities and Governments

Successful NGO work with protected areas is often a result of their acting as a bridge between communities and government. NGOs, using their local community skills and, in developing countries, drawing on expertise and funding through international links, are able to fill gaps in government-sponsored conservation programs (which are often under-resourced).

The role of liaison between local communities and government is one that NGOs are particularly skilled in. Where successful, such liaison can bring lasting benefits to ongoing management of protected areas. Sharing successes in management can increase both government and local community commitment to protected areas. Close liaison with governments helps to ensure that NGO work takes account of (and can influence) national political priorities for conservation.

In Papua, New Guinea, local and international NGOs played an active role in the negotiations over a National Forestry Action Plan—NFAP (which followed the Tropical Forests Action Plan [TFAP] for that country). The Plan has recommended that 20 percent of Papua New Guinea be allocated as national parks and World Heritage areas. The NGOs represented community views in setting conservation and forestry policy in the Plan and have established a national association of NGOs to facilitate ongoing involvement in the implementation of NFAP. They continue to monitor work on NFAP, offering advice to the government and to major multilateral funding bodies that are financing the Plan. In other countries, less successful TFAPs and NFAPs have not involved local NGOs to any significant degree.

Negotiations for Establishing Protected Areas

Campaigns by environment NGOs to establish protected areas are usually based on the biological and wilderness values of a particular area. However, their lobbying efforts for the protection of nature have been most effective when they also recognize and try to accommodate government priorities in meeting the needs of different users of the natural resources under debate. One of the most important examples of this in New Zealand was the West Coast Accord, signed in 1986 (Maruia Society, 1986). Here representatives of regional and national governments, the timber industry, and the conservation movement met to resolve the future of a bitterly debated area of forest on the South Island's West Coast. The negotiations were chaired by the government but the principal deal was struck between the timber industry and the NGOs. The West Coast forests are important to both parties and there was strong incentive for agreement to be reached. As a result of these negotiations, a package deal was agreed that led to the protection of 90 percent of the region's forests, including a new national park and a complete system of ecological reserves. The balance of 10 percent was allocated to sustainable forestry to provide a resource base for the region's timber industry. Similar agreements between conservationists and industry have been negotiated since the West Coast Accord. In New Zealand such negotiated settlements have become valuable tools for reconciling opposing land use interests.

The involvement of conservation NGOs in the resolution process for land use decisions in natural areas is critical. Their input represents the public's concern about nature conservation and, through reasoned and well-researched argument, balances the countervailing influence of industry. Equally important to the success of the negotiations in the West Coast Accord was that both the timber industry and the conservation NGOs were willing to consider each others concerns. Because of this, the two interest groups were able to develop a mutually agreeable package rather than allow government to be the sole arbiter in the process of compromise and solution funding.

Environment groups need to ensure that they retain the flexibility and pragmatism that can be the hallmark of NGOs to enable them to constructively contribute to negotiated solution-finding to environmental problems. If timber millers are to be made redundant by the creation of a protected area, for example, environment NGOs will be more effective if they incorporate solutions to this problem into their campaign to see the forest protected. Too often environment NGOs leave the difficult juggling of nature protection and human aspirations to government. The advantage of staying involved through the solution-finding process is that a strong voice for environmental defense is present when final bargaining is taking place.

Where government interest in establishing protected areas is low, or where governments lack resources to research a case for protected area establishment, these negotiations between NGOs and industry have the potential to become a vital tool for conservation.

Protected Areas Owned and Managed by Private Organizations

In many parts of the world NGOs have complete management responsibility and even ownership of protected areas. Privately owned and managed reserves are important contributors to national protected areas systems, bringing flexibility to the means by which protected areas may be established. Where purchase of land is important for its protection and government does not have the resources to buy, or where landowners are unwilling to sell to the state, land purchase and management can sometimes be achieved by private organizations. Such NGOs represent the enthusiasm of different public groups for the protection of nature with interests as diverse as safari and wildfowl hunting, research, education, or simply a love of wild places. They are an economic way of ensuring nature protection because they access a public spirit for conservation involving much voluntary effort and contributed funds. The largest private nature system in the world is owned by The Nature Conservancy in the United States, totaling 3.12 million hectares.

Keys to success by private organizations managing their own or state-owned protected areas are based on skills in specific management techniques for the species and habitats being protected. These include control of introduced species, habitat restoration, and threatened species breeding. Public money used for land purchase usually means that there must be public access to that protected area, and this in turn requires careful management to minimize degradation of the area through overuse. Wetlands are especially vulnerable in this regard. Wetlands are commonly found in privately owned protected areas partly because of the interest of wildfowl hunting organizations and partly because they are a habitat that has tended to be neglected by state protection agencies in the past. The typically slight percentage of this important habitat that is protected has been greatly dependent on the resources and commitment of private groups.

The category of privately owned protected areas includes those that are communally owned and managed by indigenous people. In Melanesia (Southwest Pacific), 85 percent of the land is communally owned by indigenous family groups. Much of this land retains its original tropical forest cover. Until recently owners kept the forests largely intact, clearing land only for subsistence gardens. Wild animals were hunted in forests, and individual trees were felled for houses and canoes. Today, as well as increased clearance for agriculture, the forests are being rapidly lost to commercial logging. Landowners consent to timber licenses over their forest because they have no other ready access to cash—a commodity greatly desired to fuel village development. There are many examples of both national and international NGO responses to this issue. One example is in Western Samoa where international NGOs have supported establishing a protected area on communally owned land at Falealupo. Indigenous landowners were reluctantly considering selling their forests to a logging company to raise funds to build a new school. NGOs helped to raise funds in three different countries to buy the village a new school in exchange for a protection covenant for 50 years over the forest. The covenant allows continued customary use of the forest by the indigenous owners. The speed and flexibility of NGO response to the problem illustrates the contribution NGOs can make to community-level needs as well as the direct links they can facilitate between developing country and wealthy country communities.

Funding and Support for NGOs

There is wide cultural, social, economic, and political diversity in NGO backgrounds, a diversity that is also apparent in the nature of NGOs' working relationships with their respective governments. In some countries NGOs are regarded with suspicion—as a source of competition for international funds, as unsilenceable critics of political policy, or as tending to amass community-based power. In these circumstances NGOs may not be invited to contribute to government policy, and decision making can take place in an atmosphere of secrecy. It is very important that efforts be made by all parties in these situations to overcome government concerns about NGO activity. NGOs need to retain open information exchange with governments, keeping their official partners fully informed about, and involved in, their activities. NGO credibility with their respective governments is helped when NGOs show a consistent quality of effort. Governments, for their part, must remain alert to the valuable resources and strengths of NGOs and seek ways to use these in the implementation of national policy. In New Zealand, environment NGOs have positions on many government-established policy institutions including parks and reserves boards, fisheries councils, and land management authorities.

Respect for NGOs and consultation with them is growing internationally. Most multilateral agencies now require project development and evaluation to involve consultation with NGOs. Many bilateral aid and development programs have given funds for entire projects to be administered by NGOs. The Maruia Society, for example, was given a three-month consultancy to conduct an ecological survey for protected areas in the Solomon Islands by the Australian National Parks and Wildlife Service. The British Overseas Development Agency has a joint funding scheme under which they provide half the costs of development projects initiated and administrated by NGOs, contributing up to half a million pounds to individual projects. Since the mid-1980s, the Canadian International Development Agency has more than doubled its funding to NGO groups, believing they are crucial to the effectiveness of Canada's aid delivery.

In developing countries major initiatives for protected areas are likely to be dependent on international funds. This means that the increasing attention being paid to NGOs by international funding sources has important implications for protected areas. A project that has NGO involvement is likely to be more attractive to the donor community because of the cost-effectiveness of NGOs and their community-based approach to development issues—an approach that has proven to be effective for protected area management. The continued effectiveness of these projects is dependent on the financial stability of the NGOs administering them. With this in mind the governments of developed countries need to ensure that the trend of support for NGOs is continued and widened so that an increasing percentage of aid funding is administered through NGOs.

Conclusion

Many NGOs around the world are effectively engaged in establishing and developing protected areas. These efforts deserve recognition from governments and international funding agencies to ensure the continuing contribution of the range of skills, dedication and innovation that NGOs have to offer. To build such confidence, NGOs for their part must ensure that their own work is of consistent high quality and that they are prepared to engage in, or at least support, their governments in the immensely difficult tasks of juggling national and community development objectives with environmental protection.

References

Gordon, Robert. 1990. The Prospects for Anthropological Tourism in Bushmanland. *Cultural Survival Quarterly* 14:(1).

Lees, Annette. 1991. *A Protected Forests System for the Solomon Islands.* Maruia Society, Nelson. 180 pp.

Maruia Society. 1986. Various membership publications including "Maruia" and "Bush Telegraph."

Snelson, D., and P. Lembuya. 1990. Protected areas: neighbors as partners—Tanzania and Kenya. In *Living with Wildlife, Wildlife Resource Management with Local Participation in Africa.* Agnes Kiss (ed.) World Bank Technical Paper Number 130.

Chapter 23

Lessons from 35 Years of Private Preserve Management in the USA: The Preserve System of the Nature Conservancy

Will Murray

Editor's Introduction:
One of the largest owners of private protected areas in the world is The Nature Conservancy in the USA, which has some 1300 private protected areas covering well over half a million hectares. As an NGO, TNC has developed a number of partnerships with governments and other nongovernmental agencies. Through 35 years of experience, TNC has learned many lessons about how partnerships can function effectively. Will Murray explains that partnerships need to involve other people, employ lessons learned along the way, plan for the long term, multiply impact with both government and private landowners, use the best available information to design and select protected areas, and involve local people at all levels. The Nature Conservancy is now applying these important lessons to new partnerships in Latin America, the Caribbean, and the Pacific.

Introduction

The Nature Conservancy (TNC) is a private, nonprofit conservation organization incorporated in 1951 for scientific and educational purposes. The mission of the Conservancy is to preserve the full array of species, communities, and ecosystems by saving the lands and waters on which their survival depends. Since 1951, TNC has completed more than 9000 land acquisitions totaling more than 2.5 million hectares. TNC owns and manages the largest private nature preserve system in the world, consisting of 1300 preserves encompassing 526,000 hectares and 1725 rare species and communities in the United States.

The Conservancy employs a direct, tripartite methodology in pursuit of its mission: Identification and mapping of rare species and communities to enable dispassionate project selection, real estate action to save lands, and preserve management (stewardship).

• *Identification* of rare species and communities and mapping of their locations is conducted by Natural Heritage Programs, a network of data centers developed by TNC, which are subsequently taken over and housed by state agencies or universities. There are Heritage Programs in all 50 U.S. states, 13 Latin American and Caribbean nations, two Pacific island nations, and three Canadian provinces. The data centers employ a standardized methodology and use Conservancy-developed software to enable data sharing and analysis. The information is used to identify critical areas in need of protection, and to establish conservation priorities on a regional, national, and global basis. Another use of the data is to help facilitate design and implementation of ecologically sound development projects, by providing information on occurrences of rare species with the hope of steering development projects away from ecologically sensitive areas.

• *Protection* activity includes acquisition of real estate through purchase, gift, easement, lease, and other traditional real estate techniques. The Conservancy works with willing sellers and donors to acquire land with uncommonly high biodiversity values for long-term protection from development. In addition to land, TNC also acquires water rights.

• *Stewardship* includes both property management and ecological management of its native species, natural communities, and ecosystems. Custodial management, including boundary posting and patrolling, fencing, trail construction and maintenance, trash removal, and visitor management is commonly conducted by a large portion of TNC's 20,000 volunteers. Ecological management, including prescribed fire, alien species control, ecological restoration, and hydrological restoration is conducted by many of the same volunteers, the Conservancy's 220 stewardship staff and collaborators from government agencies, academic institutions, and other partner organizations.

The Conservancy's preserve system is diverse. Gray Ranch, in southwestern New Mexico, is the largest at 130,000 ha, whereas Spindrift Point Preserve in California is 1.3 ha. Old-growth forests, prairies and savannas, chaparral, marshes, riparian forests, bogs, fens and swamps, beaches, islands, desert oases, and estuaries are all represented. The preserve system has a dynamic makeup, with new properties constantly entering the portfolio and existing preserves occasionally being transferred to like-minded groups for long-term management. We also manage a large number of preserves cooperatively with academic institutions, government agencies, private individuals and businesses, and other conservation groups. The Stewardship enterprise is underlain by endowments totaling $45 million.

The Conservancy refers to land management as "Stewardship" to reflect the ultimate goal of the enterprise and our management philosophy. According to the dictionary, a steward is someone who manages land for someone else. The "someone elses" for whom we manage the preserves is first and foremost the rare species and natural communities that we wish to preserve. Secondary "someone elses" are researchers, visitors, educational groups, and neighbors—in short, present and future generations. We manage for these secondary audiences only in ways that are compatible with our primary goal, to preserve the native biota, the ecological processes under which it evolved and to which it is adapted, and—so far as we are able—the potential for it to evolve in the future.

The Conservancy employs a set of organizational values that support the mission and methods. We endeavor to be collaborative, nonconfrontational, and solution-oriented, and to use direct action. We search for multiplying factors, such as partnerships and breakthroughs in information technology, to advance the mission. And we are steadfastly single-purpose, resisting temptations to stray from our mission of biodiversity preservation through habitat protection and management.

Advantages Enjoyed as an NGO

As a nongovernmental organization, The Nature Conservancy both enjoys advantages and feels constraints. The advantages are primarily in the areas of flexibility, ability to test new methods, single purpose, capacity for quick action, ability to bring resources to government, ability to create effective partnerships, capacity to build local constituencies, and ability to raise private funds.

• *Flexibility.* As with most NGOs, the Conservancy has a high degree of flexibility to undertake novel challenges. Being in the real estate business, we initiated a "Tradelands" program that accepts gifts of ecologically unimportant lands, then converts them into cash for purchase of ecologically important property.

• *Ability to test new methods.* The Conservancy successfully completed the largest feral animal control program known, on Santa Cruz Island in California. Government agencies in similar settings had been prevented from carrying out comparable feral animal control programs by animal rights activists.

- *Single purpose.* Our single-purpose focus does not allow for the abundant distractions that organizations with multiple mandates (and multiple constituencies) feel. While we do endeavor to hear the opinions and concerns of interested parties, we are not bound by the myriad constituents that agencies must serve.

- *Capacity for quick action.* The Conservancy can move quickly to acquire for a government body a parcel of land that comes suddenly on the market without having to wait for the government budgetary process. During the record cold of the winter of 1989, the Conservancy made an emergency purchase of water from the Snake River Water District #1 in Idaho to break the ice for 500 trumpeter swans who were beginning to starve due to a lack of open water in which to feed. That purchase was accomplished in 48 hours.

- *Ability to bring resources to government.* We frequently bring resources to assist government agencies, through delivery of talented volunteers to aid in land management; volunteers are more than five times more likely to volunteer for an NGO than for government. The Conservancy helped to create cost-share programs with the United States Forest Service to bring private dollars to match government funds for specific conservation projects.

- *Ability to create effective partnerships.* The Conservancy will work with any legitimate group that shares similar goals, whether the group is another conservation body, industry, government or private individuals. We are currently engaged in a research project with the Electric Power Research Institute and have Memoranda of Understanding with such organizations as the International Right of Way Association, the National Speleological Society, and the United States Bureau of Land Management.

- *Ability to secure private funds.* This is an advantage in the sense that we are not at the whim of government legislature for operating dollars or capital funds. This "advantage," as with others in this list, can also be a constraint.

Constraints We Feel as an NGO

The Nature Conservancy is constrained in three areas.

1.   *Availability of funding.* Of the $114 billion of philanthropy in the United States in 1990, only 1 percent went to conservation causes. Lack of adequate funds affects stewardship in several ways. The Conservancy owns and manages too many preserves to do an exemplary job, a trait we have in common with many government land management agencies. In general,

many Conservancy preserves are too small and too poorly buffered from harmful surrounding land uses to be entirely effective in saving a majority of native species over the long haul.

2.   *Single purpose.* The truism that in nature "everything is hitched to everything else" complicates our attempts to remain true to the mission as we have narrowly defined it. Politics and economics, two areas where we have only lightly tread, are in themselves powerful forces.

3.   *Data management.* Information is the most expensive of endeavors. Inadequate staffing has left us with preserve managers and stewardship directors who carry most of their topical information and experiential learning in their heads, unavailable to other practitioners and succeeding managers. Inadequate funding is sometimes linked to staff turnover (when salaries are impossibly low, for example), and turnover in the absence of effective data management causes severe setbacks.

Six Keys to Success

In the pursuit of brevity, I will list six keys to success as I see them. Undoubtedly there are more than six, and other people in TNC would surely have some different ideas.

1.   *Involve other people.* Partnerships with government agencies, industry representatives, local people, and other key groups are important. No one institution possesses the full complement of talents, skills, knowledge, and interests to implement land management programs. For example, The Nature Conservancy's partnerships with government agencies work because we bring resources, build constituencies, and give credit. We bring resources in the form of private funds, volunteers, information, and sometimes advocacy. We build constituencies for the agency programs by helping to publicize and draw attention to those programs. And we give credit to our partners in ways that they will appreciate—positive press, letters to supervisors and politicians, and other means.

2.   *Employ what you are learning.* Good data management is key to putting into practice the discoveries that come with managing land and the life it supports. As we learn new things about the biota and how to manage it, the task of recording those ideas and putting them into practice becomes crucial. Staff and volunteer turnover and a growing body of stewards dictate that we practice good data management. Yet good data management takes time, and altering management plans and activities can be difficult.

The Nature Conservancy has created an information management framework to capture discoveries so that they may be retrieved and used by successors. Element Stewardship Abstracts (ESAs) are compendia of the known information on management of rare species and communities and problem species. We use ESAs to summarize what is already known and to capture new information as we develop it. ESAs are contained in a database of a larger information system, the Biological and Conservation Data System, a relational database software developed by The Nature Conservancy. We also rely heavily on data management to employ new knowledge in managing alien species, in developing ecological models, and in managing individual sites with regard to neighbor relations, property upkeep, and land use history.

Employing what one learns also requires the flexibility to effect rational change. The Nature Conservancy increasingly uses conceptual ecosystem function models as a tool to record and understand ecological observations gained through management and to predict changes to the biota. Changes in management practices are grounded in the ecological model, damping erratic management shifts and providing long-term management continuity in the face of turnover, while enabling the flexibility to change as the model improves.

3.   *Plan to be involved forever.* An instantaneous act may extirpate a species or destroy a site, yet preservation takes forever. A long-term source of management funds and a meaningful, affordable monitoring scheme are necessary for permanent success.

The Nature Conservancy uses stewardship endowments to provide a long-term, dependable, and predictable source of management funds. We attempt to generate up to 50 percent of annual stewardship funds from endowments by estimating annual costs, then raising and investing sufficient funds to generate that amount on a yearly basis. Funds are invested with the simultaneous goals of long-term capital appreciation, capital preservation, and acceptable yield. We also restrict investments to socially and environmentally responsible concerns. Over the past 10 years, Conservancy stewardship endowments have grown through investments at 14 percent per annum, while we spend only 6 percent per year for management. Therefore, these funds are appreciating in real terms beyond annual spending and inflation.

Biological monitoring is our only true measure of long-term success. TNC is developing a monitoring manual in cooperation with the United States Forest Service to provide guidance to staff and volunteers on monitoring programs that are informative and affordable. A long-term presence is required for successful monitoring, and long-term stewardship success requires proper monitoring.

4. *Create multipliers.* The enormity of the task of Stewardship requires a return on our investment of time, effort, and money greater than 1 to 1. We strive to create "leverage" on our invested resources through partnerships, funding mechanisms, use of information, and engagement of volunteers. Partnerships with government land management agencies improve our ability to deliver stewardship services, such as ecological burning, weed control, and boundary patrolling. When a government partner is also our neighbor, we can share responsibilities to mutual benefit.

We frequently use funding mechanisms that stretch private dollars through government matches. The Challenge Cost Share Program of the U.S. Forest Service teams private funds with public dollars to accomplish projects of mutual interest to both parties.

Volunteers play an integral role in Stewardship. With a professional staff of 220, we simply do not have the staff capacity to manage our preserve portfolio. Neither do we have the wide range of skills and experience required in stewardship within our staff ranks. Volunteers provide sufficient numbers of people for labor-intensive programs such as restoration and visitor services, and also provide expertise in community relations, communications, engineering, scientific subdisciplines, construction, interpretation, and the thousand other needs of stewardship programs. Even if The Nature Conservancy possessed in its staff sufficient human resources to meet all stewardship needs, I would argue that we still need volunteers. Their goodwill, community support, philanthropy, media relations, advice and oversight, and staff rejuvenation activities are indispensable secondary benefits.

5. *Practice intelligent preserve selection and design.* Selection of preserves and design of preserves are the two most critical operations in land-based biodiversity conservation. The Nature Conservancy has developed and uses a standard methodology for identifying biologically significant lands. The Natural Heritage Network, a cooperative group of 135 entities using standard methodology and data management techniques, searches primary and secondary sources for information on rare species and communities and their locations, then makes these data available to legitimate users for improvement of land use decisions for both conservation and development. Most of the cooperating institutions are housed in state or national governments.

The Nature Conservancy uses information on locations of rare species and communities to drive its land acquisition programs. The Conservancy selects conservation sites on a blend of criteria including presence of concentrations of rare species and communities, quality of the occurrence, manageability, imminence of threat, and prospect of successful acquisition. Staff and volunteers draw up a prioritized list of sites that guides our land protection actions.

Preserve design is driven by three criteria, intended to preserve the target species and communities in the long term. The preserve boundaries should be drawn to capture the area occupied by the target elements, to capture enough area to protect essential ecological processes or the ability to replicate or mimic them, and should manage the property to protect it from harmful local human activities. While the professional literature blooms with theoretical information about preserve design, the managers of nature preserves are in the best position to evaluate the designs of former colleagues, as managers live with preserve design decisions every day. For this reason, the Stewardship Division reviews all preserve designs for manageability to ensure that we do not paint our future managers into a corner.

6.   *Involve local people at all levels.* Most threats to nature preserves are local. Those that are not, such as atmospheric deposition of metals, ozone depletion, and global climate change, are outside the scope and responsibility of the local manager. Involvement of local people is a key factor in nature preserve management to turn potential threats into probable successes.

The Nature Conservancy works with neighbors, local media people, local government officials, business people, and educational institutions to make sure that local communities know what we plan to do, who we are, and most important, what our vision of the future for the preserves looks like. In turn, we create avenues for local people to convey to us their concerns, aspirations, wishes, and vision of their future. By involving local people in our planning, we share that future vision.

Volunteer opportunities generate powerful opportunities for local involvement. In suburban Chicago, a 4000-person network of volunteers provides primary management for a 172-unit nature preserve system in both public and private ownership. In northern California, 5000 volunteers are actively involved in riparian forest restoration. In both cases, most of the volunteers are local people who have a personal stake in seeing the stewardship of the local preserves succeed.

In southern Arizona at the Patagonia-Sonoita Creek Preserve, Conservancy staff had long battled vandalism by local students who used the preserve as a location for loitering and other unsavory activities. The Conservancy preserve manager worked with the local judge to obtain the services of people serving a sentence of community service in lieu of jail time for minor transgressions. This year the Conservancy has benefited from more than 1000 hours of community service. More important, the word has spread throughout the community that "the preserve isn't a cool place to party anymore" as local people are now involved in cleaning up litter and vandalism. As a result, the incidents of vandalism have virtually disappeared and the stature of the Conservancy in the community has improved.

Local understanding and community involvement create and maintain the single most important factor in long-term management implementation—an active presence.

Conclusion

After almost four decades of managing nature preserves, The Nature Conservancy is learning that forces outside our boundaries exert influences on the biota that we desire to preserve. The Conservancy's newly launched Last Great Places Campaign is an attempt to work with people in the areas surrounding our preserves to minimize harmful impacts. Using a concept developed by UNESCO's Man and the Biosphere Program, the Conservancy has selected 75 sites in Latin America, the United States, the Caribbean, and the Pacific to test ecosystem conservation. The Last Great Places Campaign is working to protect biodiversity through the actions of local people.

The only meaningful biodiversity preservation is lasting preservation, and only the actions of local people can ensure that protected areas remain protected. The Stewardship of the Conservancy's preserve system, and therefore the Conservancy's mission, hinges on this effort.

Chapter 24

Partnerships Between Rural People and Protected Areas: Understanding Land Use and Natural Resource Decisions

John Schelhas and William W. Shaw

Editor's Introduction:
Partnerships between protected areas and their neighbors are often hampered by substantive conflicts over natural resources that cannot be resolved by process-oriented approaches. In these cases adjacent land uses must be promoted that meet both conservation and development needs. These land uses can be more successfully promoted by better understanding how people living around protected areas make land use and natural resource-use decisions. These decisions are influenced by socioeconomic factors such as returns to labor, availability of labor, risk, availability of cash and credit, security of land tenure, market factors, off-farm employment, knowledge of cultivation techniques, and cultural preferences. John Schelhas and William Shaw, from the University of Arizona School of Renewable Natural Resources, apply this approach for understanding land use adjacent to a national park in Costa Rica.

Introduction

Protected areas throughout the world are becoming increasingly isolated as surrounding natural habitat is converted to human-dominated land uses. This process is reducing the size of previously larger ecosystems and threatening the biological integrity of ecosystems and the survival of some critical species. Because this wildland conversion is often in part a product of people seeking to meet their basic development and survival needs, efforts have shifted from a confrontational approach between protected areas and local people to an approach that seeks to meet both conservation and development needs.

Most of the published literature in this field has stressed process-oriented approaches to resolving conflicts between protected areas and their neighbors. These approaches include public participation, informal consultation, advisory committees, management committees, collaborative problem-solving, and environmental education (Hough, 1991; Bidol and Crowfoot, 1991; Garratt, 1984). Other approaches have recognized the need to provide alternative sources of income if destructive land and natural resource uses are to be stopped, and have focused on providing direct employment to local people or park-related employment by increasing local involvement in tourist services (see Zube and Busch, 1990 for a review). Both of these approaches are important, but in many cases substantive conflicts over resources and basic development needs are not fully addressed by process-oriented approaches and cannot be adequately met by employment in the park or local tourism industry. In these cases, land uses must be promoted adjacent to protected areas that meet both conservation and development needs.

Land uses that meet both conservation and development needs can potentially include extractive uses of forest products (Allegretti, 1990), natural forest management (Buschbacher, 1990; Hartshorn, 1989), new agroecosystems such as agroforestry and intercropping that can provide sustained returns (Alcorn, 1990; Gliessman, 1988; Van Orsdol, 1987), and intensification of agriculture on existing lands to reduce the need for further wildland conversion (Schelhas, in press; Subler and Uhl, 1990). Many of these efforts have failed or achieved only limited success because, focusing disproportionately on biological conservation objectives, they have promoted land uses inappropriate for local social and economic conditions (see Chapin, 1988).

One way to improve the success of efforts to form partnerships between conservation agencies and local people is to better understand how local people make land use and natural resource use decisions. These decisions, even when environmentally destructive, are generally rational efforts to meet development needs and desires with the available resources and under the existing socioeconomic conditions (Vayda, 1983). Understanding how decisions are influenced by different socioeconomic factors enables the creation of conditions and incentives that encourage more sustainable land uses that will be readily adopted by local people.

Factors Influencing Land Use and Natural Resource Choices

The field of anthropology has produced literature on land use choice that is of considerable utility for managing land use adjacent to protected areas.

While this approach provides no easy answers, it does identify a number of key factors that should be kept in mind in the design and implementation of conservation and development strategies.

Boserup (1965) has provided a theory of agricultural intensification that has great relevance to the management of protected areas and adjacent lands. Boserup maintains that more intensive land uses produce greater return to land at the expense of declining returns to labor. Because intensification requires working harder per unit of output, people only intensify their land use when required by increasing population density. For example, shifting cultivation generally provides better returns to labor (though returns to land are generally lower) than permanent cultivation. Therefore, as long as land is plentiful, people will choose shifting cultivation over permanent cultivation.

One implication is that protected lands may be viewed as potential land for colonization, and local people may have a strong preference for expansion of agriculture because of the declining returns to labor that result from intensifying use of existing agricultural lands. The actual situation is not that simple. The choice between expansion and intensification of land use is also influenced by environmental factors and land use capability; by desires to be close to healthcare, schools, and other social amenities; and by needs to be near roads and towns for marketing products and obtaining nonagricultural employment. Yet, in all cases, returns to labor will be an important determinant of land use choice and will play a significant role in the acceptability to local people of new, alternative land uses to meet conservation and development needs.

Risk is another key factor influencing land use and natural resource use choice (Wilk, 1981; Barlett, 1982). Risk can be related to environmental uncertainty, difficulty in cultivating certain crops, unpredictable markets for cash crops, and uncertain access to resources such as cash or agricultural inputs. Many households use diversified economic strategies to minimize the risk of complete failure. This may take several forms. One is combining low risk with high risk land uses, such as combining cattle, which has low but reliable returns, with high risk cash cropping in Costa Rica (Schelhas, in press; Edelman, 1985). It may also involve using cattle or trees as savings accounts to meet family emergencies or other unexpected cash needs (Schelhas, in press; Chambers and Leach, 1990).

Security of land tenure is often considered a prime factor behind the willingness of people to invest in land improvements or land use systems that avoid land degradation in the long term. This is often the case, but several other factors also need consideration. First, de facto tenure may be more important than de jure tenure (Seymore, 1985). That is, having one's land rights respected by one's neighbors may be more important than legal title. In addition, secure land tenure does guarantee investment in land improvements that avoid land degradation or ensure that, for example, trees will be planted and cared for. Other factors, such as risk and returns to land and labor must still be taken into consideration.

Many agriculturalists in the world today, even in remote regions, use purchased agricultural inputs. The availability of cash or credit to purchase these inputs thus plays a critical role in choice of land uses, crops, and agricultural techniques (Moran, 1987; Barlett, 1982). These factors are particularly significant to small landholders who may have little available cash and limited access to credit.

Many small landholders today are becoming increasingly involved in cash cropping. Existence of markets, access to them, and their long-term stability are thus important factors behind land use choice. Improved road access adjacent to protected areas can have both positive and negative impacts on their neighbors. Improving roads can raise standards of living by permitting cash cropping, reduce reliance on park resources by permitting alternative land uses, and improve access for tourists and thus increase tourism revenue in local communities. On the other hand, roads can also make a protected area more accessible to poachers, and can lead to increased colonization or intensification of land use that increases the rate of conversion of natural habitat to human-dominated land uses.

Just as rural landholders are becoming increasingly involved in cash cropping, in many regions of the world they are also increasingly reliant on off-farm employment or nonagricultural work to meet household needs (McMillan et al., 1990; Collins, 1987; Deere and Wasserstrom, 1979). Off-farm employment can improve the standard of living of people without further conversion of natural habitat, especially when it takes place in tourism or other industries with lesser environmental impact. On the other hand, off-farm employment may also decrease the labor available to undertake land improvements and may lead to a cycle of inadequate labor investment, land degradation, and declining yields (Collins, 1987).

Knowledge of different land use techniques and options is also an important factor in land use choice. Extension programs that promote alternative land uses are often required for land use change. Along with knowledge, cultural factors and preferences exert a powerful influence on land use choice. These factors will influence the directions that land use change takes, although land use options that produce significant material benefits and are locally appropriate in terms of returns to land and labor, risk, and cash investment requirements will often rapidly overcome all but the strongest cultural resistance.

The exact influence of many of these factors in a region may be difficult to determine without extensive research. Yet considerable insight into land use choice can often be obtained by rapid assessment techniques that can be applied in less than 10 days (Schelhas, 1991). An awareness of the key factors that have been found to influence land use and natural resource choices in many different regions of the world helps the protected area manager understand the rationale behind existing land uses and the acceptability of alternative land uses for buffer zones, biosphere reserves, and multiple-use protected areas. While this approach involves only a seemingly small shift in attitude, this shift requires overcoming a common prejudice in the conservation community. Rather than viewing rural people as engaging in irrational, environmentally destructive land uses, an assumption may be made that, although the land uses may be environmentally destructive, they are probably rational given the resources and knowledge available to people engaging in these uses. The result of this change can be far-reaching. For example, rather than going into a local community seeking to convince people to plant more trees in the way that benefits the protected area the greatest, one would go into a community first seeking to understand how local people can tangibly and immediately benefit from more trees and then promoting solutions that meet the needs of both the people and the protected area (see Murray, 1986, for an example).

An Example from Costa Rica

An approach based on the above ideas was taken during research on land uses adjacent to Braulio Carrillo National Park and the La Selva Biological Station in northeastern Costa Rica, providing an example of the utility of the methodology. The northern sector of the 44,901 ha park is narrow and small. This sector also contains the only lowland habitat in the park, habitat that is important to the survival of many key park species that are seasonal altitudinal migrants. Both sides of the park are bounded by private lands, which are a part of the Central Volcanic Cordillera Biosphere Reserve. These lands are being heavily colonized and the remaining forest is rapidly being converted to pasture and agricultural lands.

Research on land use and natural resource use decisions of park neighbors sought to determine the rationale behind existing land use patterns and identify trends toward more sustainable land uses. Cattle pasture, the dominant land use in the region, is often considered one of the most environmentally destructive land uses in the tropics (Goodland, 1980). A closer look suggested that, while this may be true for large cattle ranches used primarily to hold land for speculative reasons, in Costa Rica many small cattle operations have intensified land use and achieved relatively high, sustainable stocking rates through intensive management of pasture grasses (Schelhas, in press). Animal husbandry using cattle is one of the most attractive land uses in this region due to low risk, possibility of gradual entry, market stability, and liquidity of investments in animals.

Many residents were entering into intensive cash cropping of black pepper, coffee, or other permanent crops, where this was made possible by access to markets, availability of credit, and knowledge of cropping techniques or availability of technical assistance. These permanent cash crops produced returns to land many times greater than cattle, showing potential to reduce the amount of land required to support a household and thus slow forest clearing. Many people chose to combine the high-risk cash crops with cattle, preferring diversified land use strategies that have potential to produce high income, provide insurance against crop failure and changing markets, and include a store of value to meet emergency needs.

Forest-based land uses were not often actively chosen by residents of this region, but research suggested that they would be more widely adopted if several obstacles could be overcome. Recommended strategies included (1) improved forest management technical assistance to improve the returns to land from secondary forest management, (2) agroforestry systems that combine agricultural crops or wildlife ranching with forests to overcome the low returns to land of forest and the high initial investment required by reforestation, (3) identification of fast-growing, valuable timber species for use in reforestation, and (4) increasing the amount paid to landholders for sale of standing timber (Schelhas, 1992).

A twofold strategy was recommended for managing land use on private lands adjacent to the northern sector of the park. The first component was protecting as much critical habitat as possible within the park boundary, because the value of adjacent land for conservation of the park's biological diversity was found to drop off dramatically as forest disturbance increased. The second component was promoting a mosaic of land uses, including pasture, annual crops, permanent crops, and forest, on remaining private lands adjacent to the park.

Conclusion

Effective management of lands must often resolve substantive conflicts over resources. This requires a better understanding of how local people make land use and natural resource decisions in order that strategies can be developed that both address their needs and promote solutions that will be acceptable to them. The results may at times be counterintuitive, such as promoting agricultural intensification to reduce reliance on park resources or accepting land uses such as cattle pasture that have been widely pronounced unsustainable in the lowland tropics. But accepting solutions for adjacent lands that have a high likelihood of acceptance by local people, even when these are less than ideal from a conservation point of view, may in the long run result in greater conservation gains. This does not reduce the need to adequately protect sufficient habitat in core protected areas. But if social needs are adequately met, future pressure on these areas from local people should be reduced.

Acknowledgments

Partial funding for this research was provided by a Tinker Foundation Field Research Grant and a Research Fellowship from the Jessie Smith Noyes Foundation and the Organization for Tropical Studies.

References

Alcorn, Janis B. 1990. Indigenous Agroforestry Strategies Meeting Farmers' Needs. 141–151. In Anthony B. Anderson (ed.) *Alternatives to Deforestation: Steps toward Sustainable Use of the Amazon Rain Forest.* New York: Columbia University Press.

Allegretti, Mary Helena. 1990. Extractive Reserves: An Alternative for Reconciling Development and Environmental Conservation in Amazonia. 252–264. In Anthony B. Anderson (ed.) *Alternatives to Deforestation: Steps toward Sustainable Use of the Amazon Rain Forest.* New York: Columbia University Press.

Barlett, Peggy F. 1982. *Agricultural Choice and Change: Decision-making in a Costa Rican Community.* New Brunswick: Rutgers University Press.

Bidol, Patricia, and James E. Crowfoot. 1991. Toward an Interactive Process for Siting National Parks in Developing Nations. 283–300. In *Resident Peoples and National Parks: Social Dilemmas and Strategies in International Conservation.* Tucson: University of Arizona Press.

Boserup, Ester. 1965. *The Conditions of Agricultural Growth: The Economics of Agrarian Change under Population Pressure.* New York: Aldine Publishing Company.

Buschbacher, Robert J. 1990. Natural Forest Management in the Humid Tropics: Ecological, Social, and Economic Considerations. *Ambio* 19(5):253–258.

Chambers, R., and M. Leach. 1990. Trees as Savings and Security for the Rural Poor. *UNASYLVA* 41(161):39–52.

Chapin, Mac. 1988. The Seduction of Models: Chinampa Agriculture in Mexico. *Grassroots Development* 12(1):8–17.

Collins, Jane. 1987. Labor Scarcity and Ecological Change. 19–37. In P.D. Little, M.M. Horowitz, and A.E. Nyerges (eds.) *Lands at Risk in the Third World: Local Level Perspectives.* Boulder, CO: Westview Press.

Deere, Carmen Diana, and Robert Wasserstrom. 1979. *Ingresso Familiar y Trabajo no Agricola entre los Pequehos Productores de America Latina y el Caribe.* Seminario Internacional sobre la Produccion Agropecuaria v Foregal en Zonas de Ladera en America Latina. Turrialba, CATIE, Costa Rica.

Edelman, Marc. 1985. Extensive Land Use and the Logic of the Latifundio: A Case Study in Guanacaste Province, Costa Rica. *Human Ecology* 13(2):153–185.

Garratt, Keith. 1984. The Relationship between Adjacent Lands and Protected Areas: Issues of Concern for the Protected Area Manager. 65–71. In Jeffrey A. McNeely and Kenton R. Miller (eds.) *National Parks, Conservation, and Development: The Role of Protected Areas in Sustaining Society.* Washington, D.C.: Smithsonian Institution Press.

Gliessman, Stephen R. 1988. Local Resource Systems in the Tropics: Taking the Pressure Off the Forests. 53–70. In Frank Almeda and Catherine M. Pringle (eds.) *Tropical Rainforests: Diversity and Conservation.* San Francisco, CA: Academy of Sciences.

Goodland, R.L. 1980. Environmental Ranking of Amazonian Development Projects in Brazil. *Environmental Conservation* 7(1):9–26.

Hartshorn, Gary S. 1989. Sustained Yield Management of Natural Forests: The Palcaza Production Forest. 130–138. In John 0. Browder (ed.) *Fragile Lands of Latin America: Strategies for Sustainable Development.* Boulder, CO: Westview Press.

Hough, John. 1991. Social Impact Assessment: Its Role in Protected Area Planning and Management. 274–283. In *Resident Peoples and National Parks: Social Dilemmas and Strategies in International Conservation.* Tucson: University of Arizona Press.

McMillan, Della, Jean Baptiste Nana, and Kimseyinga Savadogo. 1990. *Onchocerciasis Control Program Land Settlement Review Country Case Study.* Binghamton, NY: Institute for Development Anthropology.

Moran, Emilio F. 1987. Monitoring Fertility Degradation of Agricultural Lands in the Lowland Tropics. 69–91. In P.D. Little, M.M. Horowitz, and A.E. Nyerges (eds.) *Lands at Risk in the Third World: Local Level Perspectives.* Boulder, CO: Westview Press.

Murray, Gerald F. 1986. Seeing the Forest while Planting the Trees: An Anthropological Approach to Agroforestry in Rural Haiti. 193–226. In D.W. Brinkerhoff and J.C. Garcia Zamor (eds.) *Politics, Projects, and People: Institutional Development in Haiti.* New York: Praeger.

Schelhas, John W. 1991. A methodology for socioeconomic assessment of land use adjacent to national parks. *Environmental Conservation.* 18(4):323–330.

———. 1992. Socioeconomic and biological analysis for national park buffer zone establishment. In *Science and Management of Protected Areas*. J.H.M. Willison, S. Bondrup-Nielsen, C. Drysdale, T.B. Herman, N.W.P. Munro, and T.L. Pollock (eds.). Amsterdam, The Netherlands: Elsevier Science Publishers. 165–169.

————. In press. Building Sustainable Land Use on Existing Practices: Smallholder Land Use Mosaics in Tropical Lowland Costa Rica. *Society and Natural Resources*.

Seymore, Frances. 1985. Ten Lessons Learned from Agroforestry Projects in the Philippines. USAID. mimeo.

Subler, Scott, and Christopher Uhl. 1990. Japanese Agroforestry in Amazonia: A Case Study in Tomé-Açu, Brazil. 152–166. In Anthony B. Anderson (ed.) *Alternatives to Deforestation: Steps toward Sustainable Use of the Amazon Rain Forest*. New York: Columbia University Press.

Van Ordsdol, Karl G. 1987. *Buffer Zone Agroforestry in Tropical Forest Regions*. Washington, D.C.: Forestry Support Program, U.S. Forest Service.

Vayda, Andrew P. 1983. Progressive Contextualization: Methods for Research in Human Ecology. *Human Ecology* 11(3):265–281.

Wilk, Richard R. 1981. *Agriculture, Ecology, and Domestic Organization among the Kekchi Maya of Belize*. Ph.D. dissertation, University of Arizona.

Zube, Ervin H., and Miriam L. Busch. 1990. Park–Local Populations: An International Perspective. *Landscape and Urban Planning* 19(2):115–132.

Chapter 25 —————————————————————

Working with People Who Live in Protected Areas

Michael Dower

Editor's Introduction:
England and Wales have 11 "national parks" that are IUCN Category V protected landscapes within which some 250,000 people live. These people own most of the land within the parks and control much of the economic activity there. Michael Dower, National Park Officer at Peak National Park, describes how the National Park Authorities are expected to protect and enhance the natural beauty of the parks, provide for visitors (who number more than 100 million a year) and for the economic and social needs of residents. They have developed sophisticated means whereby the purposes of conservation and recreation are pursued in close partnerships with the residents. These means include advice and financial aid to farmers and landowners, to encourage them to manage land in a way that protects wildlife, landscape, and historic values; advice and financial aid to owners of historic buildings and other features, to encourage their sound maintenance; full recognition, in the land use planning process, of the social and economic needs of local people, plus direct action to meet those needs; encouragement of new economic activity, consistent with the conservation policies of the protected areas; and regular and intensive consultation and collaboration with organizations who work with local people. These measures, suitably modified for local conditions, could be relevant to many protected areas whose management objectives enable local populations to participate in such partnerships.

Introduction

This chapter draws its ideas and examples mainly from the experience of the 11 National Parks in England and Wales. Ten of these Parks were created in the 1950s, under the National Parks and Access to the Countryside Act 1949. The eleventh, the Broads, gained equivalent status in 1989. Between them the 11 parks cover a total area of 14,000 sq km, about 9 percent of the total land area of England and Wales. Compared with some massive national parks of North America and Australia, these areas are relatively small; but their special character comes from the fact that England and Wales have been continuously settled for many thousands of years and are now densely populated. So these national parks are not like those areas that carry the same title in the rest of the world. They are not owned or managed by the State; they are not wilderness; their use is not confined to conservation of nature, plus research and limited recreation. They are areas of mountain, moorland, enclosed farmland, woodland, wetland, and coast, long used by people for hunting, grazing, farming, timber production, and much else, and set with farm buildings, hamlets, villages, and towns. They fall within Category V of the IUCN system.

Within the 11 parks live a total of about 250,000 people, about 0.5 percent of the national total. They own most of the land within the parks: they control much of the economic activity there. They aspire to reasonable living standards.

The Challenge to the National Park Authorities

This character of the national parks poses a substantial challenge to the National Park Authorities, who by statute are given two prime functions.

(a) "to preserve and enhance the natural beauty of the areas," and

(b) "to promote their enjoyment by the public . . . "

but they are also required

(c) "to have regard to the social and economic well-being of local communities" (HMSO, 1949).

This trio of purposes has been the subject of much debate over the years among those who are concerned with the national parks. Those who fought to found the parks, in the 1930s and 1940s, perceived no inherent conflict between protection of the landscape and the well-being of the local people, and particularly of the farmers whose traditional methods had created the landscape. They saw no inherent conflict, either, between these purposes and the needs of visitors, who they thought would mainly wish to walk and relax among the mountains (Dower, 1945).

By the 1960s and 1970s, however, significant tension between the three purposes had appeared. As farming became more intensive, it began to cause radical change in the landscape. Enjoyment by the public became no longer simply walking and relaxation, but hundreds of thousands of cars coming into the quiet roads. Official reports addressed the conflict between the three purposes and emphasized that, where hard decisions had to be made, they should be in favor of the first purpose: "to preserve and enhance the natural beauty of the areas" (Sandford, 1974).

Changes of circumstance in the 1980s, and particularly changes in agriculture, have moved the debate into a more positive phase, focused on the concept of *interdependence* among the three purposes. This recognizes that most of the land in the national parks is privately owned and is used for farming, grazing, or forestry. The landscape beauty that the Park Authorities seek to protect has been created by farmers and others and is maintained by them. If farming falters, the landscape deteriorates. Similarly, maintenance of the charm of the small towns, villages, and farmsteads depends on a lively community and local economy. Visitors to the parks are served by the local community, and in turn help by their spending to support the local economy.

With this concept, the emphasis in the work of the Park Authorities is shifting away from measures designed to prevent conflict (though these measures are still needed and are taken) toward programs that seek to achieve interdependence and to realize the mutual support that landscape, recreation, and the local people can give to each other.

The Institutional Context

The National Parks are not administrative islands. They are subject to the same pattern of administration as the other parts of the UK—part of Europe, whose patterns of administration have been handed down from the forms established by the Roman Empire, based on government by sector and by hierarchy. We have separate ministries for agriculture, for environment, for employment (including tourism), for transport, and for education. We have separate national agencies for countryside (landscape) protection, for nature conservation, for sport, for tourism, and for rural development.

This sectoral approach to government can make it very difficult to pursue integrated programs of policy and action. It is also strongly marked by a downward view, whereby government sees the population as customers with broadly undifferentiated needs, rather than partners and communities with (in each locality) quite distinct needs.

The National Park Authorities are superimposed on this administrative pattern. Each of the 11 parks has its own separate authority, set up to pursue the three specific purposes described earlier. They are hybrid bodies, with governing committees or boards composed partly of nominees of the local authorities in the area, partly of nominees of the Secretary of State for the

Environment. Thus the National Park Authorities could be seen as yet another public body in an already confusing scene. But, in fact, their special constitution (with a link between central and local government) and their strong focus on a particular well-defined area, can give them a crucial ability to act as an integrating force, a bridge between the sectors of government and between government and the people.

This integrating force is powerfully expressed in one key mechanism, namely the *National Park Plan*. Since World War II, the UK has developed an effective system of land use planning. Alongside this, during the last two decades, we have developed also a new system of countryside management, of which the National Park Plan is a prime expression. This plan, prepared for each national park by its Park Authority, sets out detailed management policies to conserve the character and qualities of the park, and to provide for public enjoyment, while paying regard to the needs of the local community. These policies are negotiated and agreed with all the relevant public and private agencies, including those bodies who represent local residents. The plan provides a crucial starting-point for cooperation among the Park Authority, other agencies, and the local people.

Working with Farmers and Property Owners

The National Parks were designated in order to protect the quality of the landscape and of the built heritage of farms and villages and small towns. But this quality was not created by the Park Authorities: It was inherited by them only 40 years ago, from generations of people who farmed the land, planted the woods, and built the villages. To achieve their purposes, the Park Authorities must influence the actions of those who now own or manage the land and the buildings. Increasingly, they seek to do so not by foreseeing and preventing conflict, but by achieving interdependence.

In the 1960s and 1970s government policies tended to encourage and even to force farmers throughout the UK, including those in National Parks, to intensify their use of land in order to produce more food more efficiently. This process began to have very damaging impacts on the landscapes that the National Park Authorities were set up to protect—for example, through enclosure of heather moorland and its conversion into grassland; drainage of wetland pastures to make arable land; and use of artificial fertilizers on flower-rich upland meadows, which soon became effectively grass monocultures.

This damage led to escalating concern among National Park Authorities and to pressure on the government to halt the intensification of farming in these precious areas. Hard negotiation led to schemes whereby (first) farmers would agree to avoid further enclosure and conversion of heather moorland, in return for payments based on profits foregone (this being pioneered in Exmoor National Park); and (second) farmers would agree to maintain traditional pasture management on wetlands in return for annual payments of so much per acre (pioneered in the Boards). These schemes, which were essentially means of preventing conflict, gained national application and then wider usage in the European Community.

Events of recent years have enabled the National Park Authorities to move more fully into partnership with the farming community. In the early 1980s the success of the postwar agricultural policy moved Europe over the brink from shortage to surplus of major food commodities. Government signals to the farming community abruptly changed. Quotas were imposed in 1984 on production of milk, a major element in the livestock farming of the National Parks. More recently a sharp fall in the price of sheep, another major farm product, has coincided with a rapid rise in the public and political consciousness of the need to protect the environment, for example, from damage through intensified farming.

These two factors—the cuts in farm income and the rise in environmental concern—provide a crucial opportunity for the National Park Authorities. Farmers are looking for new sources of income; changes in food policy reduce their incentives to intensify production; environmental concerns bring increased government money into active programs to protect and enhance landscape and wildlife. In this context, all the National Park Authorities are developing "farm conservation schemes," through which they help farmers to gain income from traditional methods of farming, coupled with active enhancement of landscape and wildlife. In some parks, for example, grants are paid to farmers to rebuild or repair traditional hedges or dry-stone walls; to restore decayed woodlands or to plant new ones; to protect and enrich wetlands or flower-rich meadows. The Park Authority acts as the farmer's friend, sorting out which of a complex set of government grant schemes is best for him and "topping up" with its own grants.

A similar partnership is rapidly developing between National Park Authorities and the owners of historic buildings. The landscape of England and Wales is remarkable for its diversity of landform and of underlying rock. The special character of each National Park is strongly reinforced by the local building traditions, which use the local stone or timber and which heighten the sense of topography.

For this reason, the National Park Authorities regard the "built heritage" as an integral and significant part of the landscape that they are seeking to protect. Increasingly, they seek positive partnership with the owners of the historic buildings. They seek to prevent conflict, for example by preventing the unsuitable conversion of an historic building. But "negative" action of that kind will not save a traditional field-barn whose function as animal shelter has disappeared. Nor can all owners of historic buildings be expected to carry, unaided, the increasingly heavy cost of repairs using traditional materials such as slate or stone-slab roofs.

As a means to deal with this problem, all the Park Authorities offer expert advice to owners of historic buildings on the restoration, repair, and (where appropriate) new use of their buildings. They may act as broker to find new owners; help the owners to obtain grant-aid toward restoration work; and themselves offer such grant-aid. They may stimulate new uses, for example the conversion of redundant farm buildings into "camping barns" (Peak National Park) or "bunkhouse barns" (Yorkshire Dales National Park), which are classic examples of interdependence, producing new life for old buildings, cheap accommodation for visitors, and secondary income to the farmer.

Meeting Social and Economic Needs

A second broad area of cooperation between National Park Authorities and the people who live in the parks is in programs to meet social and economic needs. The Park Authorities all have powers of land use planning and development control. This means that they can prevent, or modify, proposed developments that would damage the landscape of the parks. Generally speaking, they do act in this way; but the obligation to "have regard to the social and economic well-being of local communities," coupled with the presence among the Park Authorities of nominees of the local authorities in the area, means that they strive to find solutions that meet the needs of local communities.

Sustaining or creating of employment within the locality is a factor that weighs heavily in the decisions of Park Authorities. The parks tend to have relatively weak local economies, with low wage levels and limited job opportunities. Schemes that offer new jobs for local people tend to gain favor with the authorities, provided they do not damage the landscape. Several of the Park Authorities are active partners in programs to strengthen the local economy, particularly where such programs also serve their primary purposes of protecting and enhancing landscape or promoting public enjoyment of the area. Examples are programs of tourism development (Exmoor and Dartmoor National Parks), the promotion of use of home-grown timber (Brecon Beacons) and of local products (Peak), and the use of local contractors in building work (Northumberland, Lake District).

Another key area of cooperation is in housing. In the rural areas of England

and Wales, house prices are pushed upward by the demands of retired people, vacationers, or commuters from the cities. Local people on low incomes may be wholly unable to afford to buy homes within the parks where they want to live or work. The National Park Authorities are all working with housing associations and other bodies to find ways to meet the need for "affordable housing."

Involving the People

Finally, the National Park Authorities are making growing efforts to involve the people in the care of their own environment and in expressing and meeting their own needs. All the Park Authorities are developing forms of liaison with, or information to, their local communities. These forms include publication of booklets and newsletters aimed at local residents, appointment of community liaison officers (as in North York Moors, Brecon Beacons, and Snowdonia), the holding of regular meetings with parish or community councils, detailed public consultation when preparing local plans, special educational activity in local primary schools, and much else. The aim of all these efforts is not merely to enable local people to understand what the Park Authority is seeking to do, but to give them the opportunity to express their own needs to the Park Authority.

Equally significant are the initiatives that some Park Authorities have taken to involve local people in action that yields environmental or social benefit. One example is the Integrated Rural Development Project in the Peak National Park, through which six public agencies pooled their funds to offer a flexible system of grant-aid to people in three parishes in the park: This prompted an upsurge of initiatives by the people, which has transformed their communities. Pembrokeshire Coast National Park has encouraged local people to take a direct responsibility for managing boat moorings and car parks, clearing litter on beaches, maintaining footpaths, managing woodlands, restoring vegetation on sand dunes, and other work. North York Moors National Park helps to finance a Small Projects Fund, which encourages local action.

Conclusion

The actions described in this chapter have their counterparts in many places around the world, including many protected areas. I claim nothing unique in them. What I do believe, however, is that this type of cooperation between protected area authorities and local residents is an absolutely vital part of the sensitive long-term management of these areas. If we in the protected area movement can help and stimulate each other to rapidly develop the concepts and techniques for such cooperation, we shall be able better to serve our own purposes and we shall have something very significant to offer to the broader world, in which the protection of our magnificent natural and cultural heritage has to be reconciled with the needs of present-day humanity.

References

Dower, John. 1945. *National Parks in England and Wales Cmnd 6378*. London: HMSO.

HMSO. 1949. *National Parks and Access to the Countryside Act*. London.

Sandford, Lord (Chairman). 1974. *Report of the National Park Policy Review Committee*. London: HMSO.

————————————————————————

Chapter 26

People and Their Participation: New Approaches to Resolving Conflicts and Promoting Cooperation

Chandra P. Gurung

Editor's Introduction:
This chapter is presented by one of the founders of a new kind of protected area in Nepal that is often considered to be the prototype of a new relationship between people and the rest of nature. Chandra Gurung describes the Annapurna Conservation Area Project, showing how this new approach to protected areas supplements that of the more conventional approaches and is thereby made more appropriate to the needs of local people. Realizing that the environmental problems facing modern society are very complex indeed, this chapter advocates determining appropriate site-specific solutions to these problems rather than a single approach to be applied everywhere. "Adaptive management" in the Annapurna Conservation Area is based on the principles of sustainability and people's participation. To help people conserve their resources, the project carries out activities in forest conservation, alternative energy, conservation education, tourism management, and community development. The local people are fully involved in the decision making from the very early stages.

Introduction

Natural resources in much of the developing world are being depleted at an alarming rate. Sadly, in most of the cases, the rural poor are blamed for the destruction of the forest without properly discerning the real reasons. It has been assumed that they are ignorant and negligent toward their natural resources. However, the destruction of forests cannot be assigned to a single factor; rather, the reasons are multifaceted. Poverty and subsistence farming, high population growth rate, overgrazing, lack of alternative resources for energy and basic necessities, and the pressure on the natural resources for fuel, fodder, food, and shelter is ever-increasing. Recently, the environmental problem has become further complicated by the growth of tourism. Hence, it is imperative that environmental problems be dealt with holistically.

In the past, problem areas were designated as national parks, displacing the local residents and prohibiting their use of resources within the parks. Since the establishment of the first national park more than one hundred years ago, the "Yellowstone Model" has been adopted in many areas where vast land and various alternative energy sources were available. However, it has not met with great success in many developing countries because of the differences in cultural, social, economic, and natural environments. As a result, various alternative approaches to protected area management have evolved in recent years that allow local residents some benefits from these protected areas.

This chapter identifies the issues and conflicts arising from establishing protected areas and other approaches to resolve these problems, using the Annapurna Conservation Area Project (ACAP) in Nepal as a case study where ecological protection and small-scale rural development programs are integrated to fulfill the basic needs of the local people while maintaining the biological and cultural diversity of the region. The ACAP approach is based on *sustainability, people's participation,* and *lami* (catalytic approach) as the three basic principles within which the conservation and rural development efforts are based.

The Issues and Conflicts

The notion of national parks embodied in the IUCN 1975 definition identifies a national park with exclusion of resident peoples from residence in and use of resources from national parks (West and Brechin, 1991). Based on such definitions alone, resident peoples have been displaced or deprived of traditional uses of park resources and left to suffer severe economic and social impacts without any documented proof that they were harming the resources of the park. The international conservation movement has perhaps too eagerly proselytized this concept in an attempt to save natural wonders, wildlife, genetic resources, and ecosystems around the globe; and developing country conservationists and resource managers have often too eagerly embraced these notions without questioning their relevance to their own economic, social, and cultural contexts. These tendencies are rooted historically in colonial systems, particularly in Africa and Asia (West and Brechin, 1991).

In the recent conservation literature, it is often presumed that indigenous peoples tend to live in harmony with the environment; modern societies, colonists, and other nonindigenous peoples seldom do (Brechin et al., 1991). The question must be answered, "Conservation for whom?" The Laxni Prasad Devkota, a Nepali poet, said, "If there are no people in the pristine nature who will admire its beauty?" Often it is argued that nature ought to be preserved in such a way that both national and international communities could enjoy its beauty to gain respite from the burden of modern civilization. But many subsistence communities in and around protected areas depend entirely on the resources found within the park for their daily needs. Hence, the debate should not be between whether biodiversity should be maintained at the cost of local residents or whether local people should be allowed to exploit the natural resources at the cost of biodiversity. Rather, conservation should demonstrate how biodiversity can be optimally maintained while fulfilling the needs of the local people.

Local residents have often been displaced when national parks were established. For example, when the Kidepo Valley (Uganda), the most important hunting area for the Ik tribe, was designated a national park, the Ik were displaced to the periphery. Once they were resettled into a sedentary life, becoming farmers instead of hunting-gatherers, many unforeseen social, cultural, and economic problems started unfolding, reaching a point where "the beaded virgins' aprons of eight to 12-year-old girls became symbols that these were proficient whores accustomed to selling their wares to passing herdsmen" (Calhoun, 1991). Similarly, establishing Gir National Park with the intention of protecting the famous Gir Lions in Gujarat, India, involved displacing the resident Maldharis tribe to the periphery of the park and completely prohibiting use of the park's resources. Although these families were provided land, healthcare services, a primary school, a drinking water supply, and agricultural implements to get them started, another problem arose. Once the Maldharis settled in the periphery, many new settlers were attracted. The Gir became an island surrounded by towns and farmland. Improvement and protection of habitats increased the wildlife numbers, but the wildlife could not be kept from the fields by the rubble walls created to protect them from domestic animals, leading to destruction of the croplands in the periphery of the park. The lions who follow these ungulates also came into conflict with the settlers. In short, as reported by one state government evaluation, the schemes of the mid-seventies might have helped the habitat (and even that is debatable), but they have not been successful in uplifting and maintaining the socioeconomic status of the Maldharis (Raval, 1991).

However, in the case of Swaziland's Mololotja National Park, a different approach was taken whereby the local authorities, particularly the king himself took the initiative to consult the local people before they were removed.

Negotiations for the national park were carried out for seven years and only after agreement was reached were the Swazi families removed. The major elements of Swazi culture and tradition, the language, the king, and the sacred beliefs were considered before the people were removed. Ancestral burial sites were kept undisturbed, and elders or family members who wished to visit the grave site were allowed to enter the park without paying the entry fee (Ntshalintshali and McGurk, 1991). The negative impacts of this approach in establishing the national park on the local residents were minimal.

The three examples cited above illustrate different degrees of social, cultural, and economic impacts in the course of establishing national parks. It shows the complexity of environmental problems in developing countries. Identifying a single factor as the cause of ecosystem depletion and trying to find a simple solution will lead from one set of problems to another.

In both the industrialized and the developing countries, various new approaches have been developed and implemented (refer to West and Brechin, 1991). Here, I would like to discuss the example of the Annapurna Conservation Area Project from Nepal, where nature conservation and human development are integrated.

Annapurna Conservation Area Project: A Case Study from Nepal

Natural and Cultural Environments

Nepal, with an area of 147,181 sq km, is a land of ecological contrast. For centuries, its landscape has been carved by numerous populations of Indo-Aryan and Mongoloid origin. Today, the country is occupied by 40 to 50 ethnic and tribal groups with distinct cultural and religious backgrounds. Nepal, sometimes known as "Shangri-La," with unexcelled beauty and cultural heritage, has its own environmental costs and dilemmas. Ninety percent of the people are subsistence farmers and more than 40 percent live below the poverty line, depending on depleted forest for fuel, fodder, timber, and other vital products.

The Annapurna region is a microcosm of Nepal. Within a short distance of about 120 km, the altitude varies from less than 1000 meters at mean sea level to 8091-meter Annapurna I, the eighth highest peak in the world. Due to its geographic features and terrain, it provides many microclimates supporting subtropical lowlands and temperate evergreen forests south of Annapurna, and finally alpine steppes and arid environments north of the Annapurna Himal. Thus it contains more than 100 species of orchids and many of Nepal's 700 medicinal plants (Sherpa, Coburn, and Gurung, 1986). This region contains excellent habitats for rare and endangered wildlife species such as snow leopard, musk deer, and a multitude of bird varieties including five of the six species of pheasants found in Nepal.

The Annapurna Conservation Area (ACA) is located about 200 km west of Kathmandu, the capital of Nepal. The ACA encircles the major peaks of the Annapurna Himal and includes the catchments of three major river sys- tems over an area of 4633 sq km. Politically, it includes two zones, five dis- tricts, and about 80 village development committees. It is also home to more than 40,000 inhabitants of different ethnic and tribal groups with various reli- gious backgrounds, mainly Hindus, Buddhists, Bon Po, Shamanists, and Ani- mists. Gurung, Magar, Thakali, and Manangi are the dominant ethnic groups; most are subsistence farmers.

The Annapurna region is also by far the most popular trekking destination in Nepal, attracting more than 36,000 international trekkers (over 62 percent of the total trekkers of Nepal) in 1989 and 1990, increasing to more than 38,000 trekkers in 1991 (ACAP, 1992). Furthermore, an average of one porter per trekker is required in the mountains, thereby providing consider- able employment.

Due to the high population growth rate, the influx of large numbers of trekkers, overgrazing, intensive agriculture, and poverty, the natural and cul- tural resources of this region are deteriorating. Unless sustainability of natural resources is addressed in their social, cultural, and natural context, the Anna- purna environment may be in jeopardy; forest management on a sustainable yield basis has been nonexistent until now.

ACAP is a pilot project implemented since December 1986 by Nepal's King Mahendra Trust for Nature Conservation (KMTNC), a nongovern- mental, nonprofit, and autonomous organization. ACAP addresses the prob- lem of maintaining a crucial link between economic development and envi- ronmental conservation. It recognizes that protection of critical habitats and long-term maintenance of biodiversity cannot be achieved without improving the economic conditions of the mountain inhabitants. Inhabitants being the focal point of every conservation effort, the project strives for a balance between nature conservation, tourism development, and human needs. Its aim is governed by the need for an ecosystem approach to maintaining the long-term integrity of the of the natural system while accommodating increased human usage, including tourism (Mishra and Sherpa, 1987).

Three Principles for Partnerships
The overall goal of the project is to conserve both natural and cultural resources for the benefit of the local people of present and future generations and by implementing rational management policies and programs. ACAP's

long-term objective is to benefit the 40,000 inhabitants by providing a viable means of enabling them to maintain control over the way their environment is used. Thus, ACAP has based its activities on three principles:

1. *Sustainability.* ACAP gives top priority to secure both financial stability and sustainable exploitation of the resources for local needs. Many foreign-assisted development projects have insufficient provisions for sustaining the developments, either by the local people or by the government, after the donor agency leaves. To be financially self-reliant once the funding from donor agencies is completed in the ACAP, an entry fee of Rs. 200 (about US$ 4.50, less than the admission fee for national parks) is levied to all the international trekkers visiting the Annapurna region. For the first time, His Majesty's Government of Nepal has allowed ACAP to collect the fee to be deposited in an endowment fund. Beginning this year, 50 percent of the annual user's fee is being utilized whereas the grants from donors are decreased by almost 50 percent. The interest from the endowment fund and income from entry fee will make ACAP financially self-reliant by 1995 to continue its basic programs. Thus, there will not be any financial burden, either to KMTNC and His Majesty's Government or to local people, even after the funding from donors terminates. A similar approach is also maintained among community development projects where the local people are either trained or provisions are made for the projects to continue. For example, a community health center in Ghandruk was founded by a Rs. 300,000 Endowment Fund in which Rs. 100,000 (US$ 2,500) and Rs. 200,000 (US$ 5,000) were contributed by the local people and ACAP respectively.

2. *People's participation.* For the long-term conservation of the Annapurna region, it was recognized that the interest of the local people and their needs must be considered first. Unless these people feel that fruits of conservation can be harvested by themselves and that the resources belong to them, the support of the local people cannot be generated and sustained. Thus the project considers the local people as the main beneficiary and includes them in the planning, decision-making, and implementing processes and delegates significant responsibilities for the management of the conservation area to them.

To carry out various conservation and community development activities within ACA, ACAP works through various management committees nominated or elected by the people themselves. One of the main committees of ACAP is called "Conservation and Development Committee (CDC)" as the main body with at least one member from each of the nine wards (lowest political unit) of the Village Development Committee (VDC). At least one female member must be elected or nominated in the CDC. The CDC has 15 members, four of whom are nominated by the Conservation Officer, representing as far as possible the various ethnic and social groups. The CDCs not only penalize those who violate the rules, but also control seasonal firewood collection and so forth. Similarly, Lodge Management Committees are responsible for improving the condition of the lodges, standardizing prices of badges of lodges, and preparing and pricing the menu. The Kerosene Depot Management Committee, Health Center Management Committee, and Drinking Water Management Committee each have their own responsibilities.

3. *Lami (Catalyst)*. His Majesty's Government of Nepal and various other national and international agencies have implemented various development and conservation projects in the region. ACAP aims to work with them in close collaboration to improve the quality of life of the people. ACAP uses grassroots methods to help villagers maintain control over their local resources as well as help to identify their immediate needs and priorities. As a result, ACAP considers itself a lami (matchmaker) that will bring together resources from outside to meet the needs of the local people.

Project Activities

Pilot activities initiated to help integrate conservation with human needs include various sustainable resource management programs and community development projects. They are discussed briefly below.

• *Forest Conservation*. To protect the existing natural forest, ACAP has established integrated forest management committees. These local committees have zoned the forest into fully protected, fuelwood collection, fodder, and timber zones. To reforest the denuded hills, ACAP has established nine forest nurseries. These nurseries are distributing more than 50,000 seedlings a year. Due to great demand from the community and individual farmers, ACAP assisted in establishing two nurseries run by individual farmers. These nurseries enable the farmers to practice agroforestry by growing fodder trees in and around their fields. Local schools, communities, and ACAP have planted seedlings at various places.

• *Alternative Energy*. Because of the high elevation of most of the Annapurna region, fuelwood needs for cooking and heating are acute. The recent influx of trekkers has further contributed to the high consumption of fuelwood. ACAP has been working on developing viable alternative energy technologies that will drastically reduce firewood consumption. The back-boiler water heater, based on displacement of water with circulatory coils, has become very popular. The system does not require additional firewood as it is fitted into the cooking stove. More than 120 lodges in the Annapurna area have so far installed this system.

ACAP has also introduced the "solar water heater and kerosene only policy." However, ACAP has realized that its alternative energy technologies cannot be based either on forest resources or on the imported fossil fuels that require hard currency and are subject to international geopolitical relations. An example of this was the closure of the kerosene depot in April 1989 when the trade embargo by India took place and Iraq invaded Kuwait in 1991. As a result, several microhydroelectricity plants are being installed to help replace kerosene and fuelwood.

• *Conservation Education.* Conservation education is considered the backbone for the success and sustenance of the project. Activities include school programs and curriculum development, mobile audiovisual and extension programs, public campaigns, study tours of the village leaders, income-generating activities for youth, and home visits. All these programs are still in the pilot stage.

• *Tourism Management.* In 1991, mote than 38,000 international trekkers visited the Annapurna area, of which over 60 percent were "Free Individual Trekkers" (FIT) who either trekked by themselves or by hiring a porter guide. Many of them are ill-prepared both physically (without appropriate clothing and equipment) and culturally (being ignorant of the local culture). Although local people earn more economic benefits from FIT than by organized trekking, their impact on the local cultural and natural environments have been largely negative.

As a result, ACAP is making every effort to educate the trekkers before they leave for the mountain by distributing brochures, a minimum-impact code, and audiovisual extension programs. The Annapurna Regional Museum and Visitor Information Center, which was recently opened in Pokhara, gateway to the Annapurna region, will become an important point in educating and preparing trekkers. In addition, ACAP intends to provide necessary health and rescue services to both trekkers and mountaineers.

• *Community Development Projects.* Most of the conservation areas are very remote. As a result, these places lack even the minimum basic facilities such as drinking water, bridges, and healthcare posts. ACAP has been helping the local people by providing both financial and technical assistance in bridge construction, trail repairs, supply of potable water, and building of schools. For most of the community development projects initiated and implemented, 50 percent of the contribution comes from the local people, ensuring a commitment and concern for the project from the local people.

To achieve a higher level of participation in conservation and to include women in decision-making processes, a special program targeting women was introduced in the beginning of 1990 by the Ghandruk Village Development Committee. The program focuses on the needs and priorities of women, bringing them into the "Conservation for Development" mainstream by providing income-generating activities. Their interest in forest conservation has been enormous.

- *Research and Training.* ACAP has undertaken several research activities on flora and fauna in the Annapurna region, including blue sheep and snow leopard studies in the Manang Valley, floristic study of the Annapurna regions, and a research needs assessment of the newly expanded area.

To upgrade the ability of the staff, training has been actively pursued. Three staff members have returned to ACAP after obtaining B.Sc. degrees in park management in New Zealand. Two more will return next year. Some of the senior staff members are being trained in the United Kingdom. Junior staff will be trained within and outside the country wherever appropriate.

Local lodge owners are also trained in the villages in conjunction with the Hotel Management Tourism and Training Center (HMTTC). The week-long training is designed so that lodge owners are trained in food preparation, sanitation, conservation, fuelwood-saving technology, menu costing, and safety and security of the trekkers.

Conclusion

Deforestation is not a new phenomenon in Nepal. It has occurred since the early 1800s when the kings granted lands in return for favors from nobility, resulting in deforestation. Then, as now, people were entirely dependent on land and were forced to clear the forests for their own use, resulting in considerable stress on the forest. Subsistence farming, coupled with a high population growth rate and growing dependency on forest resources for fuel, fodder, and timber, have led to further deforestation. Yet for centuries, the local people in the rural areas had their own management schemes to protect their forests. However, a new forest law in 1957 nationalized all the forests, thereby discouraging the traditional management schemes and encouraging deforestation. The problem has now become very complex. Extreme poverty and the increasing number of tourists have exacerbated the deterioration of the natural environment.

A national park is defined in Nepal as an "area set aside for conservation and management of natural environments including fauna, flora, and landscapes" (National Parks and Wildlife Conservation Act, 1973). It excludes the indigenous inhabitants who had settled in the area for ages. The existing law provides a long list of activities that are prohibited in the national parks and reserves—for instance, no hunting, killing, or capturing of any wild animal, cutting trees, cattle grazing, mining, quarrying, and clearing or occupying any land. During the establishment of the national parks, the indigenous inhabitants were either evicted from the area or excluded from the preparation of the management plan. In all of the mountain parks, objections were raised by the local people during their establishment (Upreti, 1985). They feared that the central government would confiscate their traditional rights for resource conservation and utilization (Sherpa, 1985). Growing population pressure as well as limited arable land resources thus forced the people to encroach into the national parks.

The conservation issues in Nepal, and in particular the Annapurna region, are inextricably linked to social and economic problems. The new conservation approach advocated by ACAP calls for the focus on human beings rather than any particular species of plants or animals. The designers of the project, Sherpa, Coburn, and Gurung (1986), pointed out that "National Park" designation was not appropriate for the Annapurna region because people were excluded. When surveyed, villagers did not respond encouragingly. Relying heavily on local participation, ACAP is establishing village conservation committees to integrate sound resource management into the traditional framework. More flexible than the national park approach, the project allows local villagers to continue to gather wood, graze animals, and even hunt as the project is designed on the multiple-use concept.

This is the first project initiated and implemented fully by Nepalese, although funding for the project comes from various donors such as WWF-USA, King Mahendra UK Trust for Nature Conservation, SNV/Nepal, and various other agencies. We feel that local participation in conservation activities is essential for building an effective partnership for conservation.

References

Annapurna Conservation Area Project. 1992. Progress Report on Entry User's Fee. Annapurna Conservation Area Project, Kathmandu.

Brechin, Steven R., et al. 1991. Resident Peoples and Protected Areas: A Framework for Inquiry. In Patrick C. West and Steven R. Brechin (eds.) Resident Peoples and National Parks: Social Dilemmas and Strategies in International. Tucson: The University of Arizona Press. 5–28.

Calhoun, John B. 1991. Plight of the Ik. In Patrick C. West and Steven R. Brechin (eds.) Resident Peoples and National Parks: Social Dilemmas and Strategies in International. Tucson: The University of Arizona Press. 55–60.

Mishra, H.R. 1982. Balancing Human Needs and Conservation in Nepal's Chitwan National Park. Ambio 11(5):246–251.

Mishra, H.R., and M.N. Sherpa. 1987. Nature Conservation and Human Needs: Conflicts or Coexistence—Nepal's Experiment with the Annapurna Conservation Area Project. Paper presented at the 4th World Wilderness Congress, September 11–18, Denver, CO.

National Parks and Wildlife Conservation Act. 1973. *National Parks and Wildlife Conservation Act, 1973.* Department of National Parks and Wildlife Conservation, Kathmandu.

Ntshalintshali, Concelia, and Carmelita McGurk. 1991. Resident Peoples and Swaziland's Malolotja National Park: A Success Story. In Patrick C. West and Steven R. Brechin (eds.) *Resident Peoples and National Parks: Social Dilemmas and Strategies in International.* Tucson: The University of Arizona Press. 61–67.

Raval, Shisir R. 1991. The Gir National Park and the Maldharis: Beyond "Setting Aside." In Patrick C. West and Steven R. Brechin (eds.) *Resident Peoples and National Parks: Social Dilemmas and Strategies in International.* Tucson: The University of Arizona Press. 68–86.

Sherpa, Lakpa Norbu. 1985. Management issues in Nepal's National Parks. In Jeffrey A. McNeely, James W. Thorsell, and Suresh Chalise (eds.) *People and Protected Areas in the Hindu Kush-Himalaya.* King Mahendra Trust for Nature Conservation and ICIMOD, Kathmandu, Nepal. 123–126.

Sherpa, Mingma Norbu, Broughton Coburn, and Chandra Prasad Gurung. 1986. *Annapurna Conservation Area, Nepal, Operational Plan.* King Mahendra Trust for Nature Conservation, Kathmandu.

Upreti, Biswas. 1985. The Park–People Interface in Nepal: Problems and New Directions. In Jeffrey A. McNeely, James W. Thorsell, and Suresh Chalise (eds.) *People and Protected Areas in the Hindu Kush-Himalaya.* King Mahendra Trust for Nature Conservation and ICIMOD, Kathmandu, Nepal. 19–24.

West, Patrick, and Steven R. Brechin (eds.). 1991. *Resident Peoples and National Parks: Social Dilemmas and Strategies in International.* Tucson: The University of Arizona Press. 61–67.

Chapter 27

Aboriginal Societies, Tourism, and Conservation: The Case of Canada's Northwest Territories

Ronald G. Seale

Editor's Introduction:

In Canada's Northwest Territories, aboriginal societies, tourism interests, and conservation agencies are increasingly finding that some of their varying goals can be achieved better through common cooperative action. As described by Ronald G. Seale, the Special Advisor on Parks Development to the Department of Economic Development and Tourism of the Government of the Northwest Territories, these interests have frequently been in conflict with one another in the past. However, it is clear that tourism is one of the few mechanisms available for transferring significant amounts of funding from the wealthier parts of society to the less well-off parts of Canada such as those in the remote parts of the Northwest Territories. As a result, some of the aboriginal peoples of the Northwest Territories are now looking to national parks to help them achieve their social, cultural, and economic goals. Where partnerships can be established among conservation agencies, tourism interests, and aboriginal peoples, benefits to the local communities can include exclusive access to resources of the protected area for subsistence purposes, training opportunities for positions in tourism and conservation, priority status in hiring programs and licensing of businesses, and compilation of traditional knowledge and heritage values of the aboriginal societies. It is also clear that such partnerships will have a much stronger basis when the legal status of the proposed protected area is first settled to the satisfaction of the aboriginal societies.

Introduction

The Northwest Territories comprise an area of some 3.4 million square kilometers, accounting for approximately one-third of the total area of Canada.

Within this vast area lives a population of just 55,000 people. A substantial proportion of that population consists of several aboriginal peoples—Inuit, Dene, and Metis. Most of these people live in small, remote communities where they maintain close ties to the biophysical resources of their homelands. In general, the living standards in these aboriginal communities are well below those in the rest of Canada.

Within the Northwest Territories, the period of contact between aboriginal peoples and the Euro-American society to the south varies from approximately 200 years to barely 50 years. For many of the aboriginal people, the shift from the nomadic life of seasonal camps to year-round residence in newly established settlements occurred only a generation ago.

Various kinds of economic activity have provided residents of small communities with entry into the wage economy. Several of these activities, however, have brought only short-term benefits. A mine, for example, closed after less than 10 years of operation. A system of defense installations provided very little employment after the initial construction phase. More often, such commercial projects imposed from the outside, with virtually no involvement of communities within the region, yielded no social or economic benefits even in the short term. On the contrary, in many instances, these developments resulted in serious social and economic dislocation for the communities concerned.

The aboriginal peoples of the Northwest Territories are now striving to achieve both formal recognition of their claims to title to northern lands and a measure of self-determination. At the same time, they are seeking to gain a higher level of economic self-sufficiency, improving their standards of living while maintaining the essential character of their traditional ways of life. In this regard, they are particularly interested in becoming involved in commercial initiatives that enable them to use to good advantage their knowledge of the land. Ideally, the scale of economic activity would also be modest, thus posing less threat to the social fabric of the small communities in question.

Tourism

Until relatively recently, much of the tourism industry of the Northwest Territories shared many of the unfavorable characteristics that distinguished other kinds of economic activity such as mining. Tourism then was typified by the fishing or hunting lodge that was planned, developed, and operated by entrepreneurs and staff from outside the region. Affluent guests flew in to an isolated lodge and left a week later with their trophies. They spent large amounts of money, but virtually none of that money found its way into the aboriginal communities of the region, nor did the visitors have any contact with those communities.

However, a different type of tourism has recently emerged in the Northwest Territories. It is known variously as adventure tourism or ecotourism. We might define a successful adventure tourism or ecotourism product as one that provides a visitor with outstanding opportunities to:

- View impressive landscapes, little altered by industrial society

- View and photograph wildlife

- Participate in wilderness recreational activities such as hiking, canoeing, and kayaking

- Contact, within mutually agreeable circumstances, traditional societies that maintain close ties to the land being visited

Within the Northwest Territories, this last component is critically important. Tourism initiatives are likely to gain community support if they provide opportunities for visitors and residents alike to come together in circumstances of mutual respect and benefit. A remote Inuit community will naturally favor an approach to tourism that is likely to engender greater understanding and appreciation of Inuit traditions and values, rather than an approach that ignores or threatens those values.

The scale of tourism initiatives is itself important. All but a handful of communities in the Northwest Territories have populations of fewer than 1500 people. A continuing deluge of tourists delivered by the busload or shipload or planeload would overwhelm such communities. Fortunately, given the climate of northern Canada, the likelihood of such inundations of tourists is rather small.

The adventure tourism and ecotourism markets are thus well-suited to the Northwest Territories. Tour groups tend to be small and of manageable size for Northern communities. These markets are generally composed of well-educated, sensitive, and relatively affluent individuals. Thus, although group sizes are small, expenditures per individual tend to be very considerable. The people attracted to adventure tourism and ecotourism are not seeking a sophisticated resort atmosphere. Instead they are likely to be very favorably impressed by the magnificent landscapes of the Northwest Territories and by the great herds of caribou, the massed thousands of seabirds and waterfowl, the spectacular mammals of the sea. These travelers are also likely to value highly the opportunity to share such experiences with residents of aboriginal communities in circumstances that ensure protection of the respect and dignity of both visitor and resident. For these reasons, both tourists and residents are responding very positively to the kind of tourism product that is becoming ever more prominent in the Northwest Territories.

Conservation

The vast expanse of the Canadian North with its impressive landscapes and relatively intact ecosystems has long been a focus of attention for several federal conservation agencies and, more recently, for territorial government departments as well. These conservation agencies have sought to achieve various kinds of protected status for lands and wildlife in an area that is little known to the great majority of Canadians, but that has important qualities of national myth.

Early conservation actions differed little from initiatives undertaken by the extractive industries with respect to involvement of indigenous populations, nor were they significantly better received. Conservation regimes were often imposed on the North with little consideration of the needs or concerns of residents of the regions in question. Today, conservation initiatives proceed only after extensive consultation with local communities and due consideration of the concerns of their residents.

National Parks

The history of national parks in the Northwest Territories stretches back almost 70 years. Wood Buffalo National Park, now a World Heritage Site, was established in 1922. Throughout much of the history of the park, nearby aboriginal communities were involved to only a limited extent in decisions affecting use of the park. Within the past generation, park personnel have made a concerted effort to include local communities in such decisions. However, it has been difficult to overcome what is perceived by local communities as decades of neglect and insensitivity. The lack of trust between local communities and the park administration continues to be a serious problem for Wood Buffalo National Park and has also had a serious impact on efforts by the Canadian Parks Service to establish new national parks in that general area of the Northwest Territories. Efforts, for example, to establish a national park in the East Arm of Great Slave Lake have sputtered on intermittently for 20 years. Prospects for success remain dim for the foreseeable future.

In other parts of the Northwest Territories, well removed from Wood Buffalo National Park, the Canadian Parks Service has been active for a much shorter period, and its efforts have been more successful—at least in the area of establishing and maintaining local community support for national park initiatives.

Auyuittuq National Park Reserve was established in 1972. Even at that time, however, park establishment entailed very little consultation with local interests. Yet despite the near absence of consultation in the early period, Auyuittuq now enjoys considerable local support, due in large part to its achievements in hiring and training personnel from the local Inuit community of Pangnirtung. Moreover, the continuation of traditional wildlife harvesting by aboriginal peoples was accepted from the outset at Auyuittuq, which was not the case at Wood Buffalo.

The favorable relationship achieved between the Auyuittuq park administration and the local community has also contributed to the favorable position adopted by the Inuit organization, Tungavik Federation of Nunavut (TFN), concerning establishment of new national parks. Within its land claim settlement area, which covers some 1.9 million square kilometers of the Eastern Canadian Arctic, TFN is strongly encouraging establishment of several new national parks as part of a land claim settlement.

The TFN position appears to be based on a belief that such parks would serve to protect large natural areas and their wildlife resources, while fully accommodating traditional aboriginal harvesting and providing welcome employment opportunities compatible with Inuit values. In short, some of the aboriginal peoples of the Northwest Territories are now looking to national parks to help them achieve their social, cultural, and economic goals.

Territorial Parks

Within the past 20 years, the territorial government of the Northwest Territories has established its own park system. The primary purposes of territorial parks are to support tourism and to provide recreational opportunities for residents of the Northwest Territories. Perhaps the most interesting territorial park initiatives with respect to aboriginal communities are two very successful historic parks in the Baffin Region of the Eastern Arctic. At Qaummaarviit Territorial Historic Park near lqaluit, the focus is on an archaeological site settled perhaps 700 to 800 years ago by the Thule people from whom today's Inuit are descended. The theme at Kekerten Territorial Historic Park near Pangnirtung is the whaling industry during the period of contact between Inuit and Euro-Americans a century ago.

There has been a high level of cooperation between the territorial government and local communities in both of these park initiatives. Each park has served to focus attention on the history of the Inuit people and thus to engender enhanced pride in the heritage of their communities.

Local involvement has also been a critical element in the planning and development of the Angmarlik Visitor Center in Pangnirtung. The center introduces visitors to the cultural history and natural history of the region, and also responds to visitor enquiries concerning such things as accommodation, guided excursions to Kekerten Park and other nearby sites, as well as outfitted trips to more distant destinations. As part of its role in interpreting the cultural history of the local Inuit, Angmarlik serves as a meeting point for community elders. The center also facilitates meetings between elders and tourists. It thus enriches the experience of visitors and at the same time helps to give community residents a positive view of the tourism industry and of tourists themselves. Tourism is therefore regarded locally as an activity that heightens understanding and appreciation of Inuit values and traditions rather than as a phenomenon that threatens those values. As well, tourism is considered beneficial in the provision of employment and business income. In many communities of the Northwest Territories, tourism facilities and services are owned by community cooperatives, thus helping to ensure that benefits generated by tourism remain within the communities.

Heritage Rivers

The importance of community involvement in the success of conservation undertakings in the Northwest Territories is clearly demonstrated by the heritage river initiatives. The Canadian Heritage Rivers System is a program operated jointly by the federal government of Canada and by the governments of those provinces and territories that have chosen to participate. A river is first nominated on the basis of its heritage and recreational values. The nominating jurisdiction must then prepare a management plan that demonstrates how the heritage values of the river will be protected. The river is designated as a Canadian Heritage River only after the river management plan has been lodged with the Canadian Heritage Rivers Board.

The Thelon and Kazan are two major rivers of the Northwest Territories that are renowned for their outstanding natural values, cultural/historical values, and recreational values. In seeking Canadian Heritage River status for these two rivers, the government of the Northwest Territories chose to adopt a somewhat different approach to the planning exercise than that used previously with respect to heritage rivers. The project was initiated by the small Inuit community of Baker Lake. A project planner employed by the government of the Northwest Territories lived in the community with his family for 18 months. There he worked closely with local organizations such as the Hunters and Trappers Association and the Elders' Society, as well as with the Hamlet Council. Use of this approach ensured that documentation of the heritage values of the two rivers reflected the perspectives of local residents and institutions, as well as of academic researchers well removed from the community.

With the planner living and working directly with them, the people of Baker Lake felt that the project was their project and that it was intended to serve their interests. The mayor of the Hamlet ultimately made part of the presentation to the Canadian Heritage Rivers Board in which the Thelon and Kazan were nominated as heritage rivers. This was a significant departure. Never before had a community played a direct role in the nomination process. The approach used reflected the wish of the people of Baker Lake that national recognition be afforded to their heritage as contained within the heritage of the two rivers. The community also wanted to see those natural and cultural heritage values protected and to share those values with others, in part by encouraging sensitive tourism. The two rivers were subsequently designated in 1990 as Canadian Heritage Rivers.

Assessment of Progress to Date

A general evaluation of progress to date on initiatives involving aboriginal peoples, tourism, and conservation agencies seems worthwhile at this point. It may be possible to apply lessons learned thus far, not only in other parts of the Northwest Territories, but also in other parts of the world where analogous circumstances exist.

Success achieved to date seems to have been dependent on the following factors.

1.   *Community Involvement.* There must be effective involvement by the aboriginal or traditional societies in all aspects of conservation and tourism planning. This may seem a truism, but its importance cannot be overemphasized. It is also worth noting that in situations in which such community involvement has been absent in the past, a much greater effort will likely be required, over a longer period of time, if long-held feelings of distrust and animosity are to be overcome.

2.   *Community Benefits.* In the course of on-going consultation with concerned communities, conservation jurisdictions must be able to point to tangible benefits that will flow to the communities from a prospective conservation initiative and associated tourism. Those benefits might include such things as the following.

- Continued and/or exclusive access to biophysical resources of the protected area for subsistence purposes

- Provision of technical and professional training opportunities relating to positions in tourism and in conservation agencies

- Priority status in hiring programs undertaken by tourism interests and conservation agencies

- Priority status in the licensing of businesses to be operated in the park or protected area

- Compilation of traditional knowledge and heritage values of the aboriginal societies by the conservation jurisdiction, for use both by the communities themselves in strengthening their societal traditions and by the conservation agency in giving to its visitors a heightened appreciation of the traditional society

3.   *Scale.* It is impossible to quantify with precision the levels of change that are manageable or acceptable. Nevertheless, the readiness with which a traditional society can adjust to changes brought about by a new conservation management regime and by associated tourism is clearly dependent in part on the scale of those changes. The changes wrought by the onset of mass tourism may utterly destroy the social and economic fabric of a traditional society. The small groups of visitors that typify adventure tourism and ecotourism can be accommodated much more readily. Control of visitor numbers and visitor use patterns may thus be in the best interests of both the traditional society and the biophysical resources that are the object of protected area status.

4.   *Ownership of Land.* Formal recognition of their claims to ownership of land is a primary objective of aboriginal or traditional societies in many parts of the world today. Experience in Canada suggests that the establishment of a protected natural area and the accommodation of associated tourism will be accepted much more readily by aboriginal societies if the legal status of the land in question is first settled to their satisfaction.

It may well be possible to advance land ownership initiatives and protected area initiatives more or less simultaneously. This is now occurring in the eastern part of the Northwest Territories. Elsewhere, however, in areas where there has been little progress on basic land ownership questions, negotiations concerning establishment of protected natural areas are making little headway. In short, the importance of the land ownership issue to traditional societies must be recognized and the question resolved at least in principle, before significant progress can be expected in the establishment of protected areas.

5.   *Sensitivity to Needs of Area Residents and Visitors.* It is obvious that the concerns of traditional societies must be sensitively and satisfactorily addressed, if a protected area and associated tourism interests are to achieve their goals over the long term. It is perhaps less obvious that the concerns of traditional societies can sometimes be dealt with in part by addressing with sensitivity the needs and concerns of visitors themselves. By and large, the persons who make up the adventure travel and ecotourism markets wish to accord every respect to the traditional societies whose homelands they visit. They wish to avoid actions that might be seen as offensive or detrimental to the interests of local residents.

In general it is a matter of considerable pride among these travelers that they are willing to have their actions and travel patterns constrained in order to reduce as much as possible their impact on both traditional societies and biophysical resources. However, since they are visiting lands whose people, wildlife, and vegetation are largely alien to them, adventure and ecotourist travelers must be informed as to what is and what is not acceptable. Having been so informed, the vast majority of these visitors will be happy to comply. They wish neither to embarrass others nor to be embarrassed themselves. They recognize too that compliance is in the interests of all concerned, and that the quality of their own experience depends in part on their compliance.

Ideally, traditional societies and conservation agencies work together to establish patterns in which visitors and local residents can come together in circumstances of mutual respect. In most instances, visitors wish to meet local residents, and local residents wish to share with visitors elements of their cultural and natural heritage that is so important to them. When visitor and resident can come together in a setting such as the Angmarlik Visitor Center, the interests of all are well served.

Conclusion

Within the Northwest Territories, the interests of aboriginal peoples, the tourism industry, and conservation agencies appear to be converging to the mutual benefit of all three. However, this mutually supportive situation has not existed for very long, nor is it characteristic of the whole of the Northwest Territories even now. If maximum benefits are to be derived from this coalescence of interests, it is incumbent on aboriginal organizations, tourism interests, and conservation agencies to build on their existing communication channels, to identify further areas of common ground, and to develop and implement additional cooperative initiatives. Given this requisite cooperative effort, considerable benefits can be realized by aboriginal communities, tourism interests, and visitors to the Northwest Territories. Furthermore, these benefits can be realized under regimes that entail the protection of outstanding natural and cultural heritage values. Finally, it is felt that it may be possible to apply some of the lessons learned in the Northwest Territories to other parts of the world in which similar circumstances exist. There, too, some of the goals of aboriginal peoples, tourism interests, and conservation agencies could be achieved through cooperative initiatives that serve the interests of all three.

Chapter 28

Stewardship: Landowners as Partners in Conservation

Kenneth W. Cox

Editor's Introduction:
Protected areas by definition are public lands, yet it is clear that many areas that are important parts of a nation's conservation program are held in private hands. Owners of private land can be important partners in conservation both as supporters of adjacent protected areas and as responsible land managers in their own right. Kenneth Cox, Executive Secretary, North American Wetlands Conservation Council (Canada), describes stewardship programs that build on the concern of local landowners. He points out that stewardship programs must be designed in ways that are appropriate to the local setting. Different kinds of landowner agreements, tax structures, and relations with government are possible, giving stewardship the flexibility to adapt to changing conditions. It is apparent that protected areas need partners in the private lands that also contain important biological resources. Building a "land ethic" among private and corporate landowners can be an important contribution to national conservation objectives. By enabling these landowners to contribute, protected area managers can help build broader public support for their activities, thereby building appropriate public attitudes that will enable protected areas to flourish.

Introduction

"It (the doctrine of private profit and public subsidy) expects subsidies to do more—and a private owner to do less—for the community than they are capable of doing. We rationalize these defects as individualism, but they imply no real respect for the landowner as an individual. They merely condone the ecological ignorance which contrasts so strongly with his precocity in mechanical things. But the final proof that it is bogus individualism lies in the fact that it leads us straight into government ownership. An orator could decry it as abject dependence on government, tolerated by the owners of a free country. I do not decry it, but I hate to see us lean on it as a solution."
A. Leopold, 1939.

The ecosystem is a mosaic of separate yet interdependent organisms, linked to each other through the evolution of time, and dependent on the land, air, and water resources that sustain it. One of the species in that evolution, Homo sapiens, in its quest for a better life, alters the many land forms and life processes that sustain it. Humankind is currently in a race to protect the ecological integrity of a great many ecoregions around the world, in order to maintain as full a range of biodiversity and economic sustainability as is possible. To maintain the integrity and biodiversity of our landscapes, we must employ a full range of mechanisms and institutional arrangements to protect our land, as any one particular method of protection alone is not sufficient. A full land stewardship program should be one of those arrangements.

Over time, and throughout a wide variety of cultures and jurisdictions, different ethics and attitudes toward use of the landscape have evolved. It is critical that within the next 10 years we survey and where necessary modify those attitudes that are detrimental to the ecosystem approach and to the maintenance of a full range of biodiversity. In the tool chest of different mechanisms and arrangements to protect land, perhaps attitudinal compliance to this philosophy is the most important goal for which to strive. For if a change in attitude does not occur in those areas where a poor land ethic exists, then ecoregion conservation will continue to be piecemeal, segregated, and unsuccessful. Fostering a full program of land stewardship will go a long way to promoting this attitudinal change.

One of the key messages that came out of the World Commission on Environment and Development's 1987 report, *Our Common Future*, was the need to "think globally, act locally." Certainly one of the primary objectives of the IVth World Congress on National Parks and Protected Areas was to share different philosophies and programs to foster global cooperation for protected area establishment and management. When we consider the second part of that phrase, acting locally, stewardship is one of the main mechanisms to achieve regional and local support and action for protected areas.

The Meaning of Stewardship

In attempting to define the word "steward" or "stewardship," words such as "looks after," "maintains," "conserves," or "manages" will be found. Thus, the term stewardship embodies two components, attitude and action; both are directed toward maintaining the ecological integrity of the land.

Stewardship programs look after the landscapes on which humanity works and plays. They are, moreover, valuable tools to educate and communicate to the landowner, the general public, and the decision maker about the importance of the natural wonders that exist on their land. A stewardship program, through working with both the individuals and their community, can begin to change the attitude of people in the population toward proper management of their own land. The enthusiasm for such a program is passed on through local, regional, and national politicians to their respective legislative bodies. Because the programs are voluntary, and in almost all cases accepted by the community, politicians like to be involved with them. This groundswell of activity and generation of goodwill is critical in order to build an atmosphere acceptable to policy and program change. For while the enthusiasm of the individual on the land is critical, so is the community acceptance of the program and the passage of legislation that institutionalizes the mechanisms and arrangements needed for proper and comprehensive support of such a stewardship program.

The second part of the definition of stewardship involves action on the land. One of the great strengths of a stewardship program is the flexibility that is connected with it. Many methods and mechanisms can be used to accomplish a range of protection from quite temporary to permanent. Figure 28-1 (Hilts *et al.*, 1990) indicates some of the parameters involved in dealing with stewardship programs, from simple landowner contacts, through man-

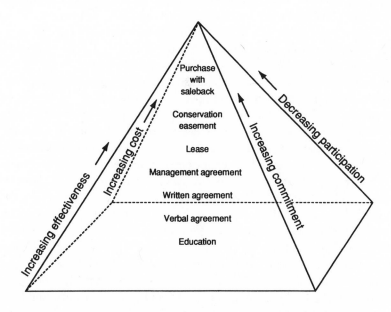

Figure 28-1. Landowner Contact

agement agreements and leasing, to more permanent conservation easements. As well as dealing with effectiveness, cost, commitment, and participation, public and private access to stewardship sites must be addressed.
Agreement by a landowner to enter into a stewardship commitment does not
necessarily bring with it increased public or private access to the land. However, in areas where private land ownership predominates in a landscape and
public access to land is limited, a public access component can be built into
the program.

As such a conservation-based attitude becomes prevalent in the community and concrete action is shown on the ground, movement toward a greater
degree of ecological integrity on the landscape will be achieved.

The Concept

For discussion purposes, stewardship can be divided into two components,
public and private. By way of example, reference will be made to the kinds of
stewardship programs that exist in Canada. Within the area of public stewardship, three categories—municipal, provincial, and federal—can be cited.
In such cases, stewardship agreements can be reached among any of the three
broad jurisdictional levels mentioned, and/or among any of these governments and nongovernment organizations or agencies. For example, while a
particular critical habitat or scenic natural area may be owned by the government, the management responsibility (stewardship) could be performed by
another government body or nongovernment organization.

The second category, private stewardship, is made up of two components:
individual land stewardship and corporate land stewardship. Individual land
stewardship programs involve agreements, be they verbal or written, with private landowners to secure and manage either wildlife habitats or areas of
scenic or natural interest. On landscapes where a significant proportion of the
land is owned by private individuals or families, and where there is little possibility of government purchasing large areas of land, maintaining species
diversity and ecological integrity on such a working landscape can only be
achieved through a dedicated stewardship program. The 1990s in much of
the world will see environmentalism, especially interest in "your own back-
yard" or local area, become an increasingly common means for individual and
community involvement in the environment. This will lead to an even
greater concern for landscapes and wildlife. Private stewardship programs will
help to make the backbone of this growing land ethic strong and stable over
the coming generations.

A second part of private stewardship involves communicating with as well as coming to agreement with corporate landholders. Many corporations hold significant amounts of land that they are either currently using or have purchased with the expectation of use in the future. Many timber, mining, and public utility companies have purchased critical wildlife habitats, in many cases wetlands, for either filling or extraction purposes and/or transportation purposes. The intention in dealing with these corporate partners is not to request funding for protected area purposes, but to communicate and educate as well as request action to manage the corporate lands that they control in an ecologically sensitive way. Many excellent examples of corporate stewardship of wildlife habitat and scenic lands exist, but many more are required. A corporate created and funded foundation called the Wildlife Habitat Enhancement Council (WHEC) provides such a service to corporate partners in the United States. Through contacting WHEC, ecological and planning expertise can be arranged so that corporate workers can enter into proper management plans for the land they either work on or control. Most corporate partners involved in such activities take great pride in their projects and consider that such projects have a positive effect on worker morale.

Types of Stewardship Programs

Many types of stewardship programs on the landscape are possible. These include site-specific, larger area, and species-specific projects. For example, in Canada, a great many small stewardship projects cover one to five hectares; these might be remnant critical wildlife habitat areas or a natural landform of scenic interest. In the southwest portion of Ontario, a zone referred to as Carolinian Canada encompasses a stewardship program in which approximately 40 individual sites and lands adjacent to these sites are part of a stewardship program that is attempting to maintain the flora and fauna of that region. Larger area stewardship programs include major timber leases—lands leased for a specific period of time from the Crown, in which nongovernment and corporate partners have agreed to manage the forest for timber and/or wood fiber production and also to manage it for sustainable wildlife habitat. Species-specific projects may include endangered flora or fauna, such as the Lake Erie Watersnake, Burrowing Owl, or Tallgrass Prairie. Such projects range from urban or urbanizing areas to recreationally developed areas to heavily worked rural landscapes.

Landowner recognition—whether it be of a government, corporation, or private landowner—is an integral part of any stewardship program. Initial contact with, acceptance, and management responsibilities of the landowner are just a beginning. It is just as critical to maintain contact throughout the year and following years to both encourage the landowner and monitor the progress of the protection plan that was developed. Recognition comes in many different forms, whether it be a periodic newsletter, a gate or fence plaque, a conservation certificate or award, or a baseball hat and handshake. Figure 28-2 illustrates a code of conservation ethics given to an individual steward, in this case under a prairie stewardship program in cooperation with the Saskatchewan Soil Conservation Association. Such tangible recognition of both responsibility and thanks goes a long way to creating and maintaining a land conservation ethic within the individual and the community.

Based on integrity, principle, and an understanding of the fragile prairie environment, a prairie steward strives to:

- Manage the land for the benefit of both present and future generations
- Harvest agricultural products on an economic yet sustainable basis
- Restore damaged or degraded lands to beneficial agricultural, wildlife habitat, or environmental uses
- Ensure that the land is protected against soil degradation
- Utilize agricultural chemicals only when necessary and in a manner that does not harm the land, environment, or the health of neighbors
- Reduce and reuse by-products of the agricultural enterprise in an environmentally sound manner
- Support activities for advancing soil conservation
- Gain a greater understanding of new and existing soil conservation techniques
- Encourage society to conserve the land resource

Figure 28-2. Code of Conservation Ethics (from Saskatchewan Soil Conservation Association, Inc.)

The Future of Stewardship

Overall, the fostering of both stewardship programs and the policy and institutional mechanisms required to maintain them, fund them, and direct them in perpetuity are critical to the development and maintenance of a jurisdiction's land use management program for an ecoregion. In an area dominated by private land (individual or corporate) ownership, a well thought out and adaptive stewardship program is integral as one of the land protection and management practices that will help to maintain the integrity of that particular region. Figure 28-3 provides a representation of a number of the kinds of programs required in any protected areas program to maintain such ecological integrity on the landscape. The park, ecological reserve, critical wildlife habitat areas, river valleys, trails or corridors, and stewardship programs must be used together to maintain ecological integrity and species diversity on the working landscape. Alone, none of these pro-

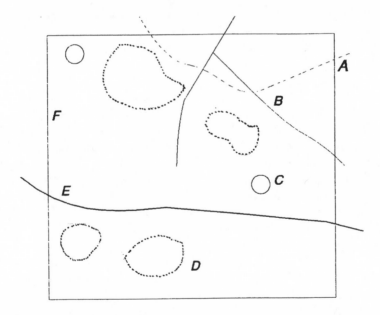

Figure 28-3. Ecoregion Protection. Sustainability of biodiversity requires a range of mechanisms. A, park boundary; B, river valley; C, ecological reserve; D, areas of private/public stewardship (protected areas not always the same, but remain a designated percentage of the jurisdictional boundary); E, corridor (trail/rail line); F, jurisdictional boundary.

grams or mechanisms can achieve the desired result, but together they provide a real possibility of achieving ecological integrity and economic stability on the land.

Private stewardship programs vary greatly in structure and delivery, but have one intent, that of land conservation securement and management. Different kinds of landowner agreements, communications plans, public profiling, local tax structures, inventive land retention mechanisms, funding, and community involvement are relevant in different settings. Flexibility and adaptability to the community is one of the great strengths of private stewardship programs.

Nevertheless, some people may have differing points of view with regard to the practicality and permanency of such stewardship programs. Critics of such programs contend that the time and effort expended would have been better spent on buying parcels of land. But isolated parcels of land do not make up a landscape, nor do they instill in people the pride and love in their immediate and local surroundings, where the vast amount of our lives are spent. They do not teach people how to be stewards of their own land. And if we are to have a rich sustainable landscape on which to live, stewardship of lands that are in private ownership is necessary.

Those interested in fostering stewardship, particularly private stewardship as part of a protected land strategy, have many examples from which to choose. Some of these examples include the Civic Trust and Groundwork Trusts, as well as the Countryside Commission from England and Wales; Scandinavia's Ecomuseums; and France's Parcs Naturel Regionaux. Heritage Canada has adapted its heritage regions approach from such European examples. Common to all of these approaches is their use of entrepreneurial innovation to tie both natural and cultural aspects of "heritage" together. Thus, through a grassroots movement, both individual and community commitment can be forged to create a community vision that will perpetuate and maintain both the ecological and economic health of a particular region.

These are just a few of the many programs currently underway that involve aspects of private land stewardship. Such examples will help to convince others that pockets of good land stewardship do exist. Such working examples will be valuable in communicating and educating others to the necessity of both a proper attitude toward land conservation and establishing new stewardship programs. Such examples also provide a range of mechanisms and institutional arrangements that can effect action for conservation on the ground.

Conclusion

Examples are only part of the story. Stewardship programs must be designed locally, regionally, or within a particular political jurisdiction, taking into account the culture, religion, history, social structure, economy, and current use of the land. They must address and work within the existing land planning framework and finance and tax arrangements of a particular jurisdiction. But because stewardship programs are both innovative and flexible, they can be adapted to a multitude of situations. It is, therefore, up to the proponent of such a stewardship program to take the best of what is offered, adapt it to a particular jurisdiction, and work to make such a stewardship component of a protected area plan an integral component of that plan.

References

Cox, K.W. 1989. *Progress and Opportunities in Advancing a Landscape Approach to Natural Heritage Conservation*. Heritage Conservation and Sustainable Development Conference, Occasional Paper 16. Heritage Resources Center, University of Waterloo. Waterloo, Ontario.

Heritage Regions. *Program Description*. 1989. Ottawa, Ontario: Heritage Canada.

Hilts, S., T. Moull, M. Van Patter, and J. Rzadki. 1990. *Natural Heritage Landowner Contact Training Manual*. The Natural Heritage League, University of Guelph. Guelph, Ontario.

Leopold, A. 1939. "The Farm Wildlife Program: A Self Scrutiny." Unpublished essay from Curt, Mine (1988) *Aldo Leopold: His Life and Work*. Madison, Wisconsin: University of Wisconsin Press.

Trombetti, O., and K.W. Cox. 1990. *Land, Law, and Wildlife Conservation: The Role and Use of Conservation Easements and Covenants in Canada*. Wildlife Habitat Canada. Ottawa, Ontario.

Chapter 29 —————————————————————————————

The South African Natural Heritage Program: A New Partnership Among Government, Landowners, and the Business Sector

Michael Cohen

Editor's Introduction:
The South African Natural Heritage Program is an innovative cooperative venture between government conservation bodies, the business sector, and the private landowner. Michael Cohen, from the Department of Environment Affairs, explains that the program is designed to complement the system of protected areas that are an important feature of South African land use. As with the stewardship programs described in the previous chapter by Ken Cox, the most important player in this partnership is the individual who at his own discretion dedicates some of his land to nature conservation, without giving up any rights to the property. The partnership is based on rather abstract incentives, especially prestige and a feeling of making a contribution to national conservation objectives. But South Africa also brings the business sector into the partnership, as a sponsor of some of the conservation measures. Further, Natural Heritage sites are exclusive, with only 151 sites qualifying to date. Other sites, which do not meet the stringent national criteria, are designated as Sites of Conservation Significance, supplemented by a Heritage Garden Program for urban settings. The three programs each have a different role, but all involve the public in conservation and help to meet local, regional, and national resource management needs.

Introduction

Southern Africa contains a remarkable physiographic diversity and richness of plant and animal species. Although relatively small, southern Africa contains some 20,000–30,000 species of vascular plants, or about 8 percent of the world's vascular flora. Among the vertebrates, 84 amphibian, 286 reptilian, 600 avian, and 227 mammalian species have indigenous breeding populations in southern Africa (Siegfried, 1989). This represents 2, 5, 7, and 6 percent of the world's species in these groups. Forty-four, 31, 6, and 15 percent of the amphibians, reptiles, birds, and mammals breeding in South Africa are endemic.

Based on an analysis of species lists for protected areas, Siegfried (1989) concluded that 34 percent of southern Africa's vascular plant species, 92 percent amphibian, 92 percent reptilian, 97 percent avian, and 93 percent mammalian species are represented in protected areas. While this may appear satisfactory, a large part of South Africa's natural diversity is threatened and the long-term survival of many of the species is uncertain. Further, while many of the species occur in protected areas, they are underrepresented in the present protected area system (Siegfried, 1989).

This chapter looks at the South African Natural Heritage Program as a means of improving nature conservation on privately owned land and as a means of involving the private sector in national nature conservation projects. It describes the evolution of the program to include the Sites of Conservation Significance Program and the proposed Heritage Garden Program which we hope will involve a large sector of the community in a total conservation effort in the urban environment.

Why a New Program?

Historically, most land that has been allocated for protection in South Africa has been purchased or acquired "in fee" by state conservation agencies. There has always been a reluctance to protect with anything less than direct purchase. This has resulted in many valuable areas being lost while the authorities held out until land could be acquired.

It is becoming more difficult to acquire land for conservation purposes. Private landowners are reluctant to sell their land. State conservation budgets are shrinking. Land values are escalating. With the current poor economic situation there is an increasing opposition to any withdrawal from the potential tax base. The past political system created an inequitable land distribution that must be rectified. Attempts to rectify the inequities of the past will place greater pressure on protected areas and on areas requiring protection. Acquisition is a slow process and too often, while the protracted negotiations are underway, the very elements requiring protection are lost or destroyed.

Given the limited and shrinking ability to acquire land, coupled to the backlog of conservation projects, it became obvious that more and new techniques for protecting our natural diversity would have to be developed. One such tool for protection is the South African Natural Heritage Program and its related sister programs (Cohen, 1985; Cohen, 1988).

The South African Natural Heritage Program

The South African Natural Heritage Program was launched on 7 November 1984, to coincide with the opening of the new Telemecanique factory in Johannesburg. The program is a cooperative venture between the private landowners on the one hand and the government sector, with the Department of Environment Affairs, the regional nature conservation agencies, the National Parks Board, the Department of Agriculture, and the private sector through Telemecanique, a telecommunications firm that was the founder sponsor, and the Southern African Nature Foundation on the other. The most important participant in the program, however, is the private landowner who dedicates a tract of his or her land to nature conservation.

The Heritage Program can best be described as citizen conservation and presents the opportunity for individuals to participate actively in the protection of our rapidly disappearing natural areas. The program aims to encourage the protection of important natural sites, large or small, in private and public ownership. By informing landowners of the special attributes of a particular site, registration as a Heritage Site reduces the chance that significant natural values may be unwittingly degraded or destroyed. It is our hope that recognition of a site will discourage others from disturbing the area. Participants are full partners and are contributing to creating a better environment for all South Africans.

The program is special in many ways. It is a voluntary program and participation is at the discretion of the landowner. In return for registering a site and the promise to protect the area, the owner of a site receives a certificate of appreciation, signed by the State President, who is patron of the program. The owner also receives a bronze plaque which indicates that the site is considered of national importance. Perhaps the greatest reward is the satisfaction that the individual will gain by voluntarily participating in a national conservation program. As the country's economy improves, we will try to negotiate a tax concession for registered landowners.

Another special feature is that the owner of a site maintains full rights over the property and can withdraw from the program by giving 60 days notice to the Department of Environment Affairs. Should the owner degrade or destroy the site, the Department will withdraw the site from registration. There are no legal attachments to the program. On registration, each site-owner receives management advice essential for the maintenance or enhancement of the site. A further advantage is that Telemecanique and the Southern African Nature Foundation have established a fund to support the program. At the launch of the program Telemecanique pledged US$ 12,000 over a five-year period and more recently have pledged a further US$ 12,000. As this fund grows, and it is imperative that it does, monies will be used to support conservation projects essential to the maintenance of elements of value found on registered sites.

From the commencement of the program it has always been Teleme-canique's philosophy to involve other sponsors. This is a national program and it is their sincere desire to see it grow. We believe that the program forms a solid base on which to build a suite of conservation projects, and as such we welcome and encourage other sponsors to join in the program as full partners or to link to the program through special projects.

In this way we have welcomed the participation of

- Hulett Aluminum, who participate by providing material for road signs that are erected on a registered site at the discretion of the owner.
- FBC Holdings, who provide expert advice and herbicides for the erad-ication of alien invasive vegetation on registered sites.
- Sage, who provided funds for the program.
- ISM, who produced their 1991 calendar on the important topic of bio-logical diversity and linked it to the Heritage Program.

The achievements of the program are not only in the elements of value that are being protected. Through the program we are communicating on a one-to-one basis with a growing segment of the population. Here we do not only include registered site owners, but also members of the general public who are interested in the program. A newsletter is produced by Teleme-canique and the Department of Environment Affairs and mailed to all inter-ested parties. For a period of two years Telemecanique sponsored an article of conservation interest every two months in a national farming journal. This was undertaken in the early phase of the program to promote the program as well as general conservation awareness.

The press, national and regional radio and television have given good cov-erage to the program. Various women's journals, teenage magazines, and a wide spectrum of educational and environmental journals have all given support to the program.

During 1989, the Department of Environment Affairs linked a broad-spectrum environmental education project to the South African Natural Heritage Program. Heritage 100 was launched to celebrate the registration of the 100th South African Natural Heritage Site. Through Heritage 100 we presented a truly national opportunity for all South Africans, as individuals, or through business associations or voluntary groups, to play a role in learning about and helping to preserve our natural and cultural heritage. As part of this project the South African Natural Heritage family produced a booklet, "100 Things to Do for Heritage 100." This project was well received by the South African public; more than 18,000 Heritage 100 booklets were distributed in just over two months. It is our intention to run with Heritage 100 until we register the 500th Heritage Site, when we will launch a new project.

At the inception of the Program, Telemecanique launched the now-prestigious Environmental Conservation Research Award. This award has over the past few years attracted entries from the country's leading research workers. Telemecanique has further broadened the base of the program by instituting an annual Grant-in-Aid to site owners who require funds for management work on a registered site. A cash grant of US$ 800 is given to the two most deserving requests.

Only the most significant areas of the country qualify for registration as South African Natural Heritage Sites. To qualify for registration, an area must contain one or more of the following natural values.

- Stands of special plant communities
- Good examples of aquatic habitats
- Sensitive catchment areas
- Habitats of threatened species
- Outstanding natural features

The criteria have been kept broad so that as many potential participants as possible can be recruited for the program.

As the Heritage Program progressed it was realized that with the high standards being applied to South African Natural Heritage Sites, we were excluding many other valuable sites. While these sites did not meet the national criteria, they contained elements of importance to a particular region. To overcome this shortcoming, the South African Nature Conservation Authorities launched in March 1989 a Site of Conservation Significance Program. The program is sponsored by FBC Holdings, and allows us to give recognition to the efforts of more site owners without compromising the high standards set for the national sites.

The Growth of the Program

Progress with the Heritage Program has been good. In 1985 the first 16 sites were registered. By 1991 a total of 151 sites were registered. Registered sites vary in size from 1 ha to 34,483 ha, and a total of 216,332 ha have been registered. Sixteen of the registered sites are in the ownership of state and/or statutory bodies, 13 are in the ownership of large commercial companies, and the remaining 122 sites are in the ownership of private individuals, giving a good spread of sites among the various sectors.

The elements protected by the registered sites are diverse and include numerous rare and threatened plant and animal species. One site, the Kogelbeen Cave in the northern Cape, hosts approximately 60,000 bats of six different species, one of which, Dent's horseshoe bat *Rhinolophus denti*, was until recently believed to be extinct. The Rooipoort site protects 14 rare and threatened bird species and about 4000 world-famous rock engravings depicting 55 animal species.

The Vogelgat site in the southwestern Cape contains 700 plant species, including 13 rare species and 12 species endemic to this site. A group of eight landowners in the Orange Free State have jointly registered their properties in order to give protection to the Cape Vulture.

A site in the eastern Cape contains the oldest known remains of anatomically modern man as well as the earliest evidence of the systematic use of marine resources by humans anywhere in the world. This site has recently been accorded the status of a national monument. A site owned by the Mlambo Tribal Authority in Kangwane protects the second largest of the five known populations of the rare cycad, *Encephalartos lebomboensis*.

The participants in the Heritage Program have resolved to strengthen the suite of projects currently being run under the South African Natural Heritage Program. We are planning a seminar and workshop for site owners where ways of strengthening the program will be discussed. Aspects such as legislation and better management will be addressed. The decision to tie a particular site to legislation will come from the individual landowner. The Heritage Program is a cooperative venture and all decisions will be made by the partners. A site owner's handbook is currently being produced. This will address various aspects of resource management. We have started a Heritage Hotline for rapid problem solving. Constant communication with the landowners is an essential element of a voluntary project.

Heritage Garden Program

The South African Natural Heritage and Sites of Conservation Significance Programs have served to indicate that if individuals are properly mandated and informed they will participate actively in conservation programs. No conservation program will work unless it has the support of the people for whom it was created in the first place.

The programs showed us that conservation is people business. People want to become involved and, once involved, will do a good job. It also showed us that nature conservation is not the exclusive preserve of the State. It is the individual, the group, and, in the end, the community that will make conservation work.

The lessons learned showed that we were on the right track and that we would have to broaden the base for participation. We were not providing opportunities for a sufficiently wide participation and for the most part had excluded the urban environment and its population from participation in conservation projects. By 1990 South Africa was almost 57 percent urbanized (Clarke, 1991), and this figure will grow rapidly over the next few years.

South Africa needs a broad-based conservation program that will present the opportunity for more people to become involved, to make their own decisions, and to create the type of environment they want. Heritage Gardens is being designed to meet these needs.

The Heritage Garden Project will not be prescriptive; there will be no rules, no legislation, and no fixed activities. The program will attempt to activate people to shape their urban places in an environmentally responsible manner. It will provide guidelines and suggestions for people to influence the fabric of their urban environment so that they will want to live there, and having made the choice will benefit from living there.

Through Heritage Gardens we will present guidelines. The decision to implement the actions must come from the people concerned. We will try to establish a network of expertise and the information needed to set up successful community projects. Groups and successful projects will link to others, and in this way we will build a national program. More important, we will build an environmentally aware community—a community that will interact and communicate. The long-term goal is the uplifting of the quality of life of all South Africans.

Heritage Gardens is being put into place to support the individual's or group's ingenuity. It will take ideas generated by groups or individuals to other groups. They will be able to tap into the expertise developed in various sectors of the community. In this way we will build a network of interlinked and interacting projects that will all contribute to a sustainable South African society. With this interaction will come a better understanding of the various perceptions we all have of the environment, of our varying needs, aspirations, and desires.

The Heritage Garden program will be made up of a number of elements. It will look at indigenous gardening, which through the road and pavement network will link to the open space system, thus creating a more viable biological system. It will also encourage urban agriculture and promote projects such as home vegetable gardens, "town farms," planting fruit trees in gardens and on pavements, the creation of parks, and the revitalization of derelict land that will provide opportunities for recreation and education, and at the same time contribute to the biological diversity of the urban complex. We will promote urban woodlots, recycling projects, composting. Built into all the projects will be the opportunity to generate income and job opportunities. Communities will be able to become more self-sufficient and to improve their quality of life.

Conclusion

The programs described above have contributed to developing a series of protection tools that allow conservation authorities to act quickly and that provide adequate protection to a specific site. We have also provided the opportunity for public involvement in conservation projects.

Through the three programs, we have expanded from the traditional approach to protected areas. This approach, which we believe is inadequate, was locked into the protection of a fixed or at least a minimum percentage of land. The 10-percent approach will not meet the country's demands for recreation, education, genetic resource conservation, and the many other goods and services that can be provided from protected areas and well-managed natural areas.

We will have achieved adequate conservation only when 100 percent of South Africa is under some form of conservation management. We are not pleading for a large nature reserve, but for a system where all development will be based on sound conservation principles and where people are able to participate in the decision-making process and are in a position to contribute to the shaping of their own environment.

We hope that these programs will contribute to the establishment of a national land use plan and a solid land use ethic. If we can achieve this, much more than the current 5.6 percent of the country would enjoy some form of conservation management and we would move much closer to the 100-percent ideal where all people would benefit and development would be sustainable. The people and environment conflict would dissolve and we would fill our role of people in the environment. With a proper integration of conservation and development, South Africa would be able to sustain more head of livestock, more farmers, more industry, and more people with a higher quality of life than we are currently doing.

Acknowledgments

I am truly grateful to the Department of Environment Affairs and the Telemecanique/ Schneider Group (SA) for making my attendance at the Congress possible. Sincere thanks to Maritz Wahl for all his assistance throughout the production of this chapter. To all in the Department's Media Section, my sincere thanks for their support and assistance.

References

Acocks, J.P.H. 1988. Veld Types of South Africa. *Memoirs of the Botanical Survey of South Africa* 57:1–146.

Clarke, J. 1991. *Back to Earth: South Africa's Environmental Challenges.* Southern Book Publishers. Halfway House. 332 pp.

Cohen, M. 1985. *The South African Natural Heritage Program.* Department of Environment Affairs. Pretoria.

Cohen, M., and G.I. Cowan. 1988. *The Southern African Plan for Nature Conservation and Related Programs.* Paper presented to the conference "Conserving Biotic Diversity in Southern Africa." Cape Town, 13–15 June 1988.

Rutherford, M.C., and R.H. Westfall. 1986. Biomes of Southern Africa—an objective categorization. *Memoirs of the Botanical Survey of South Africa* 54:1–98.9.

Siegfried, W.R. 1989. Preservation of species in southern African nature reserves. In B.J. Huntley (ed.) *Biotic Diversity in Southern Africa.* Cape Town: Oxford University Press.

Wahl, M., and M. Cohen. 1992. *Register of protected areas in South Africa.* Unpublished report. Department of Environment Affairs. Pretoria.

Chapter 30

Agreements Between Conservation Agencies and Tribal Neighbors in South Africa

J.L. Anderson

Editor's Introduction:
Jeremy Anderson, from the KaNgwane Parks Board, describes the history of protected areas in South Africa, pointing out that they have been developed very much on a traditional colonial model. This model was based on efficient law enforcement within the boundaries of the protected area, significant budgetary allocations from the State, and is supported primarily from the better-educated and more affluent sector of South African society. As has happened in many parts of the world, this model has resulted in alienating the people living in surrounding lands, who perceive little value in conservation as a form of land use. With the changing political situation in South Africa, the colonial approach to conservation is beginning to change, and many of South Africa's 12 conservation agencies are beginning to form much closer working partnerships with their neighbors in an effort to build stronger local support. Agreements with local people have been reached to include profit sharing, resource utilization, preservation of traditional ceremonies, and the creation of local employment. But such efforts do not guarantee success, as many problems still remain, including distrust or lack of understanding by the local communities of the motives of the conservation agency, conflict between stakeholder groupings within the neighboring communities, and conflict between neighboring communities themselves. An understanding of the limits of such agreements is important when trying to ensure an appropriate balance between national and local interests in protected areas.

Introduction

Most of South Africa's major conservation areas border or are surrounded by Tribal Areas. On one side of the fence are well-managed ecosystems with excellent wildlife populations and abundant natural resources enjoyed by largely affluent citizens and foreigners. Across the fence are communities that eke out an existence characterized by high levels of illiteracy, unemployment, and social disruption. Here, resentment toward the conservation areas is the order of the day (Buthelezi, 1991).

As conservation in South Africa will have to adapt to its rapidly changing political conditions, more attention must be given to the widely echoed call for park neighbors to derive tangible benefits from the conservation areas in their midst. Before doing this, we must examine whether

(a) such benefits can be provided on a meaningful and sustained basis,

(b) the provision of such benefits will have the expected positive reaction, and

(c) conflicting interest groups will accept that the utilization of wildlife in Conservation Areas is for the benefit of conservation and the local people.

Existing Agreements

Twelve conservation agencies in South Africa have tribal neighbors to the estate under their control. These are the National Parks Board, the Natal Parks Board, and the 10 ethnically based "Independent" and "National States" or homelands. The latter are Bophuthatswana, Ciskei, Gazankulu, KaNgwane, KwaNdebele, KwaZulu, Lebowa, Qwa Qwa, Transkei, and Venda. The agreements fall into broad categories (Table 30-1), with considerable variation in how each agreement is applied. Perhaps what is most surprising is that those agencies with no tribal neighbor agreements are all in National or Independent States.

Some of the more innovative forms of agreement warrant further discussion.

• *Land lease agreements.* In KaNgwane, the site of the tourist lodge in the Mthethomusha Game Reserve is leased to the private sector, the lease being paid into a trust fund to be used for community projects selected by the Tribal Authority. A more enterprising lease is that now applied in the 1620 sq km Richtersveld National Park, where the land on which the park lies is still owned and occupied by a Nama community. It is leased from the Nama for a period of 30 years with rights to graze an agreed number of livestock and to the controlled harvest of natural products. The lease payments are paid into a trust appointed by the community to manage the funds.

• *Revenue sharing.* Perhaps the most generous of the revenue sharing agreements are those in KwaZulu and Lebowa. In the former, 20 percent of gross

Table 30-1. The Diversity of Conservation Agency/Tribal Neighbor Agreements Among 12 Agencies in South Africa

Type of agreement	Number	Percentage
Land lease agreements	2	16.7
Share of revenue	6	50
Collection of firewood	3	25
Harvest of thatch grass	4	33.3
Harvest of medicinal plants	2	16.7
Meat—subsidized or free	5	41.7
Livestock grazing	1	8.3
Access agreements	5	41.7
No agreements	5	41.7

revenue for a reserve is allocated to the neighboring Tribal Authority; in the latter, 50 percent. In Bophuthatswana, 10 percent of gross revenue is allocated.

• *Fuelwood.* In rural areas of South Africa firewood is the primary source of heating and the per capita consumption is approximately one ton per year (Gandar, 1983). Only KaNgwane and Bophuthatswana have a regular fuelwood supply program. In KaNgwane, people from the neighboring villages are permitted to remove one head-load of fuelwood per week. The wood has to be from fallen trees, or from areas shortly scheduled for burning as part of the range management program. Currently removals from one 7500-ha reserve are 28 tons per annum, a level that is easily sustainable. People from more distant villages may collect fuelwood only for ceremonial purposes, provided they have the permission of the local Tribal Authority. In Bophuthatswana, fuelwood removal is permitted for the "kgotla" fire, which must always be kept burning.

• *Thatching grass.* In all Natal Parks Board reserves where thatching grass is abundant, tribal neighbors can harvest this on a one-for-one basis (for every bundle they take, they must cut one for the Parks Board). In KaNgwane, thatch grass is made available at no cost in designated areas. The villagers in the immediate vicinity have priority claim to this grass; persons from neighboring villages wishing to harvest this grass must have prior approval from the local Tribal Authority.

• *Harvest of medicinal plants.* This is permitted only in Bophuthatswana and KaNgwane. Tribal herbalists or "inyangas" are permitted to collect plants or plant parts under the control of staff. This seemingly minor concession is highly valued as it involves an extremely powerful section of the community.

• *Meat—subsidized or free.* In KwaZulu, game meat from the management program in Ndumu Game Reserve is sold at a subsidized price to local tribes people. In Bophuthatswana meat from the management program initially was sold at a subsidized price. However, it was found that most of this meat was purchased by local businessmen who had their own transport. This practice has subsequently been changed, and meat is now donated to local functions and worthy causes. In KaNgwane's Mthethomusha and Mahushe Shongwe Game Reserves, each neighboring village is allocated an annual quota of carcasses from the management program. These are provided at no cost for functions selected by the village.

• *Livestock grazing.* In only one reserve in the Ciskei, where the people gave their land for conservation, is livestock grazing permitted. This takes place for five months a year over the winter months and is restricted to cattle belonging to the tribe that provided the land. With the current severe drought in southern Africa, we can anticipate a disaster situation in the livestock in the tribal areas as these are normally stocked at Caughley's (1979) "ecological carrying capacity." In KaNgwane we have therefore resolved to select areas where grass can be cut for fodder by our immediate neighbors, but we will not permit livestock to be brought into the reserves. (This decision has been rescinded and a limited number of breeding cows and heifers were permitted to graze in the reserve until the advent of the rainy season.)

• *Access agreements.* Several of the organizations provide access to tribal neighbors or citizens at a reduced rate. However, the Natal Parks Board has taken the step of allowing immediate tribal neighbors free access on proof of identity and residence. In specific Natal Parks Board reserves, free access on regular routes is permitted to tribal neighbors who need to traverse the area. All agencies permit free entrance to organized school groups.

• *Employment opportunity.* Only Bophuthatswana and KaNgwane specify that their immediate tribal neighbors will receive priority in all employment opportunities. Probably the most tangible benefit conservation areas bring to their tribal neighbors is employment. The Kruger National Parks employs about 3000 local people, but on the other hand it is estimated that by the year 2010 there will be 400,000 unemployed people living along its boundaries. Therefore the park should be looking at ways of increasing employment opportunities within and adjacent to its borders.

The Problems

The Approach

Our very recent evolution of resource utilization by tribal people has developed with the premise that we, the conservators, have a goal for an area. To achieve our goal we have to make concessions to the local people; therefore we plan and manage for our objective conceding some forms of resource use to the people. In this we frequently make the error of being both prescriptive and paternalistic. Neither attitude is constructive in working toward a mutually agreed goal. If we are to have success over a wider scale, perhaps we should be looking at the issue from the opposite perspective. We could find out what the people want in the development of an area, and then plan toward ensuring that those elements important to us are conserved within the framework of the people's goal.

The Resource Base

While the obvious needs of tribal neighbors are recognized, it is important to evaluate the likelihood of a conservation area being able to satisfy any of these needs. Surveys conducted on this aspect in KwaZulu (Enfield, 1986), Bophuthatswana (Keenan, 1984), and KaNgwane (Odendaal, 1991) indicate that in the areas surveyed the number of people and their needs are so great that alone natural resources of the areas would never be able to provide a sustainable and meaningful benefit to everyone. This situation probably applies to all but the most remote areas, and certainly with the rapidly increasing human population the gap between supply and demand will widen and become universal.

The Increasing Populations

While Huntley *et al.* (1989) predict that the trend in mass urbanization will result in reduced population in the rural areas by the turn of the century, our experience in KaNgwane is contradictory. With an annual increase of 2.7 percent in the "Third World" segment of South Africa's population, we continue to experience increased settlement and demand for resources.

Tribal Neighbor Attitude

In the studies of Enfield (1986) and Keenan (1984) the majority of people were passively hostile toward the conservation area as they saw no positive influence on their lifestyles. In the studies of Odendaal (1991) there was little antipathy but again no great positive attitude. In all reserves, employment opportunity was the greatest benefit cited. Benefits were generally seen as being material, and Enfield's comment that "preservationist policies have convinced the majority of the local communities that the central complex (Hluhluwe and Umfolozi Game Reserves) is worthless from a practical point of view" is a telling one. There is no evidence to suggest that these findings are not general.

The Financial Aspects
Only a minority of the conservation areas in South Africa can be regarded as being financially viable. They rely largely on an annual allocation of funds from respective governments. Therefore, when a significant percentage of revenue is allocated to tribal neighbors it amounts to these people receiving an indirect handout from government; in effect the areas are being rented by the State. Given the sweeping political and economic changes facing the country, it is logical to assume that conservation agencies cannot expect to receive better budgets. Whether areas can continue to be maintained with a significant percentage of revenue being committed to tribal neighbors remains to be seen.

Conservation in Africa is largely shaped by economics (Leader-Williams, 1992); and before considering future options the likely socioeconomic situation in the future should be considered. It is with the following consequences in mind we should plan and develop our options for the future.

(a) The economic situation for conservation agencies is not likely to improve. Budget allocations from government will not keep pace with growth or inflation, so we will have to earn our own keep.

(b) The density of tribal people living adjacent to conservation areas will increase, as will their needs and the deterioration of their environment.

(c) The land demands of a newly enfranchised community will inevitably receive attention from politicians and government.

Conservation Agency Policy
Our older conservation agencies evolved very much after traditional Northern Hemisphere models. Their enjoyment has been, and still is, primarily by the more affluent white section of our community. In their development no policies were required to address the problems of the tribal neighbors as these people were the responsibility of other state departments.

When the National and Independent States created their own Conservation Agencies, policies and legislation were adopted almost directly from those of the older traditional bodies. Also the staff employed to develop these organizations had been schooled in the traditional molds. Only a few have developed new conservation agencies with policies attempting to address the people's needs.

Interest Group Conflict

Our experience is that the major conflicting interest groups within the communities are cattle owners versus non-cattle owners, and different political power blocs within the community. Cattle owners are a minority but they are powerful in terms of their wealth and seniority. This group tends to oppose much of the conservation development we propose. In contrast is the growing number of young people who see themselves as never having the opportunity to own cattle but that the development of conservation areas and tourism may bring them employment. In negotiating with such a community it is difficult to remain impartial.

Of growing concern is the increasing influence of the "anti-hunting" and "animal rights" movements in the North. Where the optimal use of wildlife populations may involve cropping or trophy hunting, the impact of these groups on the major markets can have a severe impact on viability of wildlife management as a form of land use in tribal areas. At home, the attitude of the wealthy citizens to the prospect of resource utilization as a policy of National Parks and Game Reserves has yet to be evaluated.

Where to Now?

Our approach to communicating with our tribal neighbors has to be improved. We can best do this with tribal people and regular contact with individuals, liaison committees, tribal authorities, and at senior government and ministerial level.

While we can plan for wider scope in resource use within the goals of each Conservation Area, we know that we cannot meet the demands. The only resource capable of increasing sufficiently to meet these demands is the foreign tourist *Homo sapiens superbus*. We will have to plan for greater use of our conservation areas by tourists, especially where tribal neighbors can have employment and derive economic benefits. Proposals to this end have been made by Enfield (1986). This could involve the creation of facilities on the park boundary within the tribal area in which private sector, conservation agency, and tribal neighbors participate. Conservation agencies should actively pursue the option of joint projects in the tribal areas. Hitherto, conservation agencies have confined their activities to within their borders. However, the situation across the fence is probably going to be the greatest factor affecting our conservation goals. Therefore, we should be proactive in these areas in trying to ameliorate conditions. We can with very little cost assist with skills training, improving the productivity of the subsistence gardeners and farmers, and guiding neighbors into productive home industries.

The experience of those agencies that have implemented resource utilization for the tribal neighbors is that the symbolic value of the programs to the people are very high, despite the fact that the quantity of resource available may be limited.

Wildlife, if reintroduced into tribal areas and managed for production, could become a new communal resource. Unlike Zimbabwe, where substantial wildlife populations have enabled the development of their CAMPFIRE program based on resource utilization (Anon, 1990), the resource in most tribal areas in South Africa is severely depleted. We do however have wildlife populations in conservation areas from which we can provide the founder populations to develop projects similar to CAMPFIRE in our Tribal Areas. We also have the ability to develop these areas with the cooperation of the people.

The experience of the comparatively modest developments in Bophuthatswana, Lebowa, and KaNgwane is embarrassing in that the demand by tribes to develop similar Tribal Resource areas cannot be accommodated for lack of funds.

Conclusion

The Conservation Agency/Tribal Neighbor interface in South Africa will become a major consideration for the future management of our Conservation Areas. Agencies will have to recognize this and become proactive in working toward identifying and resolving the anticipated problems before they arise. While the resources in Conservation Areas may never be able to satisfy the demand, sharing these with Tribal Neighbors on even a symbolic scale creates a more positive attitude toward the area. Employment opportunity is the major need, and fostering this is probably going to be where most can be achieved.

While education is seen as the key to employment, and some Environment Education programs contribute to basic education, opportunities must still be created by the Reserves if a positive attitude is to be nurtured and sustained.

Acknowledgments
Many of the views expressed in this chapter stem from discussions with K. Lane, R. Collinson, and A. van Wyk.

References

Anon, 1990. *People, wildlife and natural resources—the Campfire approach to rural development in Zimbabwe*. The Zimbabwe Trust, Harare. 25 pp.

Archer, F. 1991. Richtersveld Parks off to a shaky start. *New Ground*. 5:47.

Buthelezi, M.G. 1991. Extracts from opening address in *National Parks, Nature Reserves, and Neighbors*. Johannesburg: Endangered Wildlife Trust. 5–6.

Caughley, G. 1979. What is this thing called carrying capacity? In M.S. Boyce and L.D. Hayden-Wing (eds.) *North American Elk: Ecology, Behavior, and Management*. Laramie: University of Wyoming Press.

Enfield, M.M. 1986. *The viability of establishing industries in KwaZulu based on wildlife products from the Central Complex*. Report to the KwaZulu Bureau of Natural Resources. Pietermarftzburg: Institute of Natural Resources. 298 pp.

Gandar, M.V. 1983. *Wood as a source of fuel in South Africa*. Monograph 3. Institute of Natural Resources. University of Natal. 28 pp.

Huntley, B., R. Siegfried, and C. Sunter. 1989. *South African Environments into the 21st Century*. Human and Rousseau. Tafelberg. 127 pp.

Keenan, J. 1984. Report on the socioeconomic effects of the Pilanesberg Game Reserve on the surrounding populations, and the attitudes of the population to the Game Reserve. Submitted to the Bophuthatswana Parks Board, Dept. Sec. Anthropology. Wits. 2 Vols.

Koch, E. 1991. Mandela goes green. *Weekly Mail*. 5 April 1991: 1–2.

Leader-Williams, N. 1992. *The effectiveness of different approaches to elephant conservation in African range states*. African Wildlife Foundation, Elephant and Ivory Information Service. Nairobi, Kenya.

Odendaal, A. 1991. *A social impact evaluation for the management of the Mthethomusha Game Reserve as a Tribal Resource Area in KaNgwane*. Report to the KaNgwane Parks Corp. 112 pp.

Chapter 31 ————————————————————————

Communities, Parks, and Regional Planning: A Co-Management Strategy Based on the Zimbabwean Experience

Simon C. Metcalfe

Editor's Introduction:
One of the most innovative partnership programs in Africa is based on traditional approaches to conservation. Through the empowerment of local government to manage and administer the wildlife resource, the Department of National Parks and Wildlife Management in Zimbabwe has established the basis for co-management between themselves, local authorities, and rural communities. Principles for the Communal Area Management Program for Indigenous Resources (CAMP) have been developed into the well-known CAMPFIRE program which enables the rekindling of a communal ethic for land stewardship. The new resource tenure system in the communal sector allows the department to transform its role from that of protector to that of facilitator, devolving the authority for resource management, as well as the benefits and the costs, directly to the community. This co-management strategy has been developed to reconcile the human dimension in resource management, soften the hard edges between protected and communal lands, and develop new connections between people and resources in the landscape. As Simon Metcalfe, from the Zimbabwe Trust and Campfire Association, points out in this review, community involvement needs to empower groups with the right of access to definite resources, in bounded spaces, institutionally established and integrated into a regional planning framework. While this model will not work everywhere, its lessons are widely relevant.

Introduction

A "community" in precolonial Zimbabwe consisted of a hierarchy of land communities nesting one within the other and with membership depending on acceptance by traditional authority at each level of the hierarchy. Common resources such as wildlife, grazing, firewood, and water were regulated within this structure (Holleman, 1969). While population densities were far less and fragmentation of habitat hardly a great anxiety in the 19th century, Cousins (1987) states the land tenure system functioned as a mechanism of social control.

Interventions by the European colonial powers in the 20th century in Africa had a radical impact on traditional land tenure systems. Nations were established that cross-cut cultural and natural systems. Statutory laws were promulgated that alienated local people from land, grazing, forest, and wildlife resources. Rural people lost access to wildlands as protected areas were established and also lost legal access to wildlife on their own land. The appropriation by the State of natural resources generally led to elements of an "open access" system developing, with individual entrepreneurship invading the commons, as a collective sense of proprietorship was lost (Murphree and Cumming, 1990).

The Post-Colonial Dilemma

The independent states of sub-Saharan Africa have largely attempted to maintain the principle of state control of the wildlife estate. In doing so they have served the tenets of conventional western wildlife conservation and learned little from the rural development experiences occurring all around the protected areas.

Consequently, the parks and the park management became irrelevant to the local development effort. It is axiomatic that a management system that depends on external sanctions and incentives for success will collapse unless it is maintained. It would be a fair appraisal to state that protected areas in rural Africa protected by conventional law enforcement methods have failed, as species become increasingly threatened and habitat more isolated.

Many factors—population growth, poverty, corruption, lack of representation, nonviable local market economies, and poor planning—combine with weak law enforcement and contribute to the fragmentation of ecosystems around protected areas. If these islands were closely linked to local land use planning, on the other hand, they could contribute to local development, and local appreciation of the value of biodiversity might be greatly enhanced.

The concerns raised by conservation biologists regarding minimum viable areas for maintaining species diversity indicate that most protected areas are inadequate for the preservation of many large mammals and predators because the habitat is insufficiently stable to adequately protect biological processes and species (Wilcox, 1980; Soulé, 1985; Shaffer, 1987; Western, 1989; Cumming, 1990).

Finally, most wildlife and its habitat actually exist outside of protected areas where the policy of state proprietorship in African savanna areas has nullified the potential use wildlife could have on communal and private land.

Empowering Local Management

The fortress mentality inherent in the process of constantly defending protected areas in a losing battle against rural development led Zimbabwean ecologists to rethink wildlife management policy. It was noted that as long as all wildlife remained in the realm of the state, the public (private or communal) could not invest in it. Unless the public were granted access to wildlife there would be no possibility of a multispecies production system approach to land use. Unless some complementarity could be established between the protected and communal land, the "hard edge" approach would continue with increasing conflict over land use, management, and regional planning.

The Zimbabwe government, as the responsible authority, reconciled this conflict by granting the right to manage and administer wildlife to landowners (private) and landholders (communal) (Parks and Wildlife Act 1975). The results were immediately dramatic in the private sector, where land allocated to wildlife has expanded, rapidly competing with and complementing cattle-based range management systems (Child, 1991). In the communal sector the obvious difficulties inherent in common property resource management of a fugitive resource have been addressed through the Communal Area Management Program for Indigenous Resources (CAMPFIRE) policy, now an integral part of the National Conservation Strategy.

In Zimbabwe, the communal areas largely surround the protected areas and, consequently, largely depend on the Department of National Parks and Wildlife Management (DNPWLM) and CAMPFIRE to reconcile parks and communities within a wider regional plan.

Managing the Communal Wildlife Areas

It has become clear in the CAMPFIRE program that to achieve a positive co-management structure, three main stakeholders have to agree on an effective institutional framework. The communities, their local authorities, and the

Ministry of Environment (national parks, forests, and natural resources agencies) need to collaborate closely. The communities are the primary stakeholders, the environment agency is a technical support agency and arbitrator, and the local authority is the lowest level of formal statutory accountability.

As statutory law can only delegate powers to the lowest statutory body, the issue of accountability between local authority and communities is vital. Are the people accountable to the Council or vice versa? Which way does democracy actually work? Are the people shareholders or labor? To answer these questions, the CAMPFIRE program has established a set of principles to guide the relationship between local councils and people. These principles attempt to avoid an unfair bureaucratic tax on the wildlife resource that domestic resources (livestock) do not have to suffer.

The CAMPFIRE principles include the following (Murphree, 1991):

- Effective management of wildlife is best achieved by giving it focused value for those who live with it.

- Differential inputs may result in differential outputs.

- There must be a positive correlation between quality of management and the magnitude of benefit.

- The unit of proprietorship should be the unit of production, management, and benefit.

- The unit of proprietorship should be as small as practicable, within ecological and sociopolitical constraints.

The CAMPFIRE program cannot claim to have achieved its objectives but can claim to be positively establishing the framework for developing the local community institutional capacity for managing wildlife resources. That capacity, within acceptable community organization, will serve the management of natural resources in a holistic way as much as the single resource of wildlife. The most positive way to educate a community on the importance of the natural processes of the ecosystem is to first empower the people with responsibility for its costs and rents.

The CAMPFIRE program argues strongly in favor of sustainable use being the springboard for large mammal diversity integration into communal land use practices. A wise use approach outside a protected area is effectively subsidized by the preservationist approach of the protected zone. The park may be perceived as an ecobank supplying interest in the form of a renewable supply of wild animals. The softening of the "hard edge" between zones has always been a goal of buffer zone approaches, but too often the relationship between people and park has been asymmetrical and not a genuine meeting of land uses and authorities (Brown, 1991).

People, Wildlife, and Property

Unlike crops and domestic livestock, wildlife is a common-property resource like rivers, grazing lands, and forest areas. Wildlife is a mobile (fugitive) resource, and consequently communities need to know their boundaries and form collaborative associations with their neighbors. This is particularly necessary with large mammals and predators whose range is greater than any one basic CAMPFIRE unit.

By definition, common-property resources are ones for which exclusion of interloping appropriators is difficult. Privatization of the African rangelands is often not feasible, and state control has proved inadequate. A successful approach to this commons dilemma may be found in complementary and compatible relationships among the resource, the technology for its exploitation, the property-rights regime, and the larger set of institutional arrangements (Berkes *et al.*, 1989). In cases of communities and their relations to protected areas, cooperative management arrangements (co-management) are needed, involving the sharing of power between governments and local communities. On communal land, a common-property management system is implicit. Ostrom (1990) suggests a set of design principles that have been illustrated by examples of long-enduring common-property resource (CPR) institutions.

    (a) *Clearly defined boundaries.* Individuals and households who have rights to withdraw resource units from the CPR must be clearly defined, as must the boundaries of the CPR itself.

    (b) *Congruence between appropriation and provision rules and local conditions.* Appropriation rules restricting time, place, technology, and/or quantity of resource units are related to local conditions and to provision rules requiting labor, material, and/or money.

    (c) *Collective choice arrangements.* Most individuals affected by the operational rules can participate in modifying the operational rules.

    (d) *Monitoring.* Monitors who actively audit CPR conditions and appropriator behavior are accountable to the appropriators or are the appropriators.

    (e) *Graduated sanctions.* Appropriators who violate operational rules are assessed graduated sanctions (depending on the seriousness and context of the offense) by other appropriators, by officials accountable to these appropriators, or by both.

    (f) *Conflict resolution mechanisms.* Appropriators and their officials have rapid access to low-cost local arenas to resolve conflicts among appropriators or between them and officials.

(g) *Recognition of minimal rights to organize.* The rights of appropriators to devise their own institutions are not challenged by external government authorities.

For CPRs like CAMPFIRE that are parts of larger systems:

(h) *Nested enterprises.* Appropriation, provision, monitoring, enforcement, conflict resolution, and governance activities are organized in multiple layers of nested enterprises.

Recognition by African governments of minimal rights of local communities to organize and define their own institutions for natural resource and wildlife management is the fundamental policy principle that inhibits co-management possibilities at present. The success or failure of common-property resource management has to do with the exclusion and regulation of joint use (Berkes *et al.*, 1989). The chances of success for local level management depend critically on legitimization and support by central government. This support is lacking in sub-Saharan Africa at present. No amount of lip service to community involvement will substitute for the need to empower groups with the "right of access" to definite resources, in bounded spaces, institutionally established and integrated into a regional planning framework.

Local Institutional Development

Levels of decision making from the individual, household, village, community (local), on to district and provincial level (regional) and further to national and international levels need to be nested and united in common purpose. The benefits inherent in maintaining diversity and stability in ecosystem conservation should not be the object of conflict, particularly between higher and lower levels in the decision-making process.

Rather than a hierarchy, the decision and action process should be seen from an individual perspective of concentric circles. Each circle of participation represents an institution (household, village, community, district) that embodies some kind of collective action (Uphoff, 1986). The benefits of the institutions are "public goods," and it is suggested that the natural resource base should be perceived as the natural capital of the economic dimension of the institution.

In this conceptual framework the protected area asserts a force primarily for the local "public good" but also balances local interests with those of a wider public as well as future generations. Only through the evolution of viable, local, natural resource management institutions will the greater public, through its agencies, be able to establish a network of co-management institutions capable of not only planning but managing a wider landscape than the park.

A way exists to maintain Africa's splendid large-mammal diversity in savanna land use by facilitating the establishment of local wildlife management systems, linked with the government protected areas. Governments should not only discuss "tradeoffs" outside the park but should realize the necessity for tradeoffs inside as well. The buffer zone is not outside the park but between the perceptions of central government and local people regarding appropriate use of local resources.

Sustainable Use

Most consumptive use of wildlife in Africa is defined by governments as illegal. While efforts to protect endangered species with force are important and heroic at times, they should not hide a profound malady of the approach. While all common-property resources need protection from illegal appropriation, the loss of Africa's elephants and rhinos in the past decades are symptomatic of a massive divide in perception of value between governments and local people.

Although Western governments may believe they can protect the remnants of their wildlife diversity by large investment in enforceable regulations, it is unlikely that Africa can invest the management cost. Wildlife has to save itself, and the experience of Zimbabwe and other countries indicates that it can do so, provided it is used wisely, marketed effectively, and the rents are appropriated to the land they came from. It is a gross tragedy that elephants and rhinos, for example, have paid so much and received so little protection in return.

An African conservation ethic is necessary, and it should be based on local proprietorship and sustainable use with protected areas providing a local subsidy to ensure harmony between protected and public lands. This in turn will require governments to empower their rural people at the expense of urban people. Some communities will come off better than others. CAMPFIRE does not argue for equity but for socioeconomic justice. Those who pay the costs of having wildlife on their land must receive the economic benefit. There must be an incentive for having wildlife instead of goats and cows on the range, and it must be competitive or complementary. As McNeely (1988) states, "Behavior affecting maintenance of biological diversity can best be changed by providing new approaches to conservation which alter peoples perceptions of what behavior is in their self-interest."

McNeely goes on to say that unfortunately too little biological diversity will be conserved by market forces alone, and that effective government intervention is required. What exactly is "effective government intervention?" How it will be paid for and why it is lacking in much of Africa are issues generally skirted. How are conservation costs to be met and by whom if not by the people on the land and the countries they live in?

Respect and care for the community of life entails, among other things, an improvement in the quality of human life, the conservation of the earth's vitality and diversity, and the sustainable use of renewable resources (IUCN, 1991).

Conclusion: A Caring Partnership

The maintenance of biological diversity depends on the integration of social, biological, and economic factors. The future requires a new approach to basic needs that encompasses physical and emotional human needs as well as the needs of the ecosystem that sustains both. Governments must help their rural people get back in direct touch with the natural resources in their areas and, on the basis of unequivocal local proprietorship, begin to reestablish a true spirit of stewardship. That spirit must translate into the process of institution building for wildlife and natural resource management.

The role of the state is to facilitate this process by ensuring an enabling framework and professional technical inputs. The role of science is to support the planning, training, monitoring, and evaluation phases of the policy process. The nature of the economic system for management is critical; it should be determined locally and not bureaucratically to ensure the full impact of the incentive structure. Investments must be made into developing the human institutions and understanding the ecosystem, its resources, and their use and marketing. Rents from the resource must be returned to the community and to the land through wise and gentle management. Managing the local environment must reward itself, and communities can determine and negotiate the rates for reinvestment (capital), sustainability (recurrent income), community development, levies, and taxes.

The post-colonial synthesis proposed seeks to reunite old and new, local and central, cultural and natural diversity, in the context of modern Africa. This requires a relaxation of the right to associate and make a living without threatening central authorities. To enable co-management strategies, governments should be persuaded that the proposal is positive to their own interests. The affective economy, described by Hyden (1983), of networks of land, kinship, and support in much of rural Africa still offers a high human quality of life to many. Rural quality of life could be lifted further by forest and wildlife departments moving from being protectors to being enablers of sustainable conservation and development.

Rural wildlife or natural resource cooperatives could be allowed to negotiate with the national and international private sector related to marketing the sustainable yield from their common system. The possibility exists of regional and national associations for natural resource products supporting primary producers, able to capture the best possible values for their resources. These resources are in some cases more valuable than others and include beautiful landscapes, lake shores, rivers, and forests, as well as valuable minerals (gold, gemstones, and so on). In other areas resources are stressed by overuse. CAMPFIRE argues that you cannot generally subsidize one area from another. Spectacular resources like Victoria Falls contradict this view but reinforce the point that governments through their protected areas, forests, and powers can wield immense influence on local regional land use practices. Government only needs to empower people to manage their own resources and use the protected areas to supply further benefits in order to be a lead stakeholder in the policymaking for landscape planning. As it stands the natural resource departments in communal lands are far behind other agencies in extension services. Wildlife conservationists must participate in the development process, armed with the liberating effect of devolving resource benefits and sound technical advice as to how to sustain the benefit flow.

References

Berkes, D. Feeny, B.J. McCay, and J.M. Acheson. 1989. The benefits of the commons. *Nature* Vol. 340:91–93.

Brown, M. 1991. Buffer Zone Management in Africa. Workshop organized by PVONGO NRMS Project, QEP, Uganda. 1990. *Report WWF USA.* 97 pp.

Child, B. 1991. Wildlife Use on Zimbabwe's Rangelands. In *Developing World Agriculture, Animal Production and Health.* Hong Kong: Grosvenor Press International.

Cousins, B. 1987. *A survey of current grazing schemes in the communal lands of Zimbabwe.* Center for Applied Social Sciences (CASS). University of Zimbabwe (UZ).

Cumming, D.H.M. 1990. *Wildlife Conservation in African Parks: Progress, problems, and prescriptions.* WWF Multispecies Production Systems Project. Project paper No. 15. Harare, Zimbabwe. 20 pp.

Holleman, J.F. 1969. *Chief, Council, and Commissioner.* Royal van Gorcum. Assen.

Hyden, G. 1983. *No short cuts to progress. African development management in perspective.* University of California Press. 223 pp.

IUCN, UNEP, WWF. 1991. *Caring for the Earth. A Strategy for sustainable living.* Gland, Switzerland. 228 pp.

Martin, R.B. 1986. *CAMPFIRE—The Communal Area Management Program for Indigenous Resources.* Branch of Terrestrial Ecology, Department of National Parks and Wildlife Management, Ministry of Environment. South Africa.

McNeely, J.A. 1988. *Economics and biological diversity: Developing and using economic incentives to conserve biological resources.* Gland, Switzerland: IUCN. 236 pp.

Murphree, M.W. 1991. *Communities as institutions for resource management directors.* CASS, UZ. Harare, Zimbabwe. 21 pp.

Murphree, M.W., and D.H.M. Cumming. 1990. *Savanna land use: Policy and practice in Zimbabwe.* CASS/WVVT Multispecies project. UZ, Harare.

Ostrom, E. 1990. *Governing the Commons. The evolution of institutions for collective action.* Indiana University: CUP. 280 pp.

Shaffer, Mark. 1987. Minimum viable populations: coping with uncertainty. In *Viable Populations for Conservation.* Michael Soulé, New York: CUP.

Soulé, Michael. 1985. What is conservation biology? *Bioscience* 35(11):727–734.

Uphoff N. 1986. *Local Institutional Development: An analytical sourcebook with cases.* For the Rural Development Committee, Cornell University: Kumarian Press.

Western, D., and Mary Pearl. 1989. *Conservation for the Twenty-First Century.* New York: Oxford University Press. 365 pp.

Wilcox, Bruce A. 1980. Insular conservation and conservation. In *Conservation biology: an evolutionary-ecological perspective.* (M.E. Soulé and B.A. Wilcox, eds.). New York: Sinauer Associates Inc.

Chapter 32

Neighbors as Partners of Protected Areas

Deborah Snelson

Editor's Introduction:
The African Wildlife Foundation has developed an important new program for working with neighbors as partners of protected areas in East Africa, with demonstration
projects in Kenya, Tanzania, and Uganda. In developing this program with local partners, AWF has also provided a bridge to the government management authorities in
these regions. The lessons have been somewhat different from those of The Nature
Conservancy in the Western Hemisphere, though with a number of similarities.
AWF's experience has underlined the importance of local participation built on community self-interest in rural development. Deborah Snelson, Deputy Director of
African Operations for AWF, underlines the importance of choosing appropriate target groups, defining clear mandates, promoting self-reliance, and developing alternative approaches to wildlife use. However, real participatory involvement of local people can only show tangible results when government policy facilitates revenue sharing
with the local communities. Partnership must always be a two-way street linking governments to local communities, but NGOs can often serve as facilitators.

Introduction

The importance of wildlife authorities taking the initiative in developing contact with communities around the areas under their jurisdiction has been
demonstrated in a few pilot projects in Africa—most notably the Communal
Areas Management Program for Indigenous Resources (CAMPFIRE) in Zimbabwe and the Lupande Development Project in Zambia, which was
expanded into the national program Administrative Management Design
(ADMADE). The term "community conservation" is used by the African
Wildlife Foundation (AWF) to describe the approach of AWF's "Protected
Areas: Neighbors as Partners" program in East Africa. The principle that

local communities should be involved in and benefit from conservation of protected areas is now accepted, but there is still comparatively little experience of how to put the principle into practice. AWF is pioneering this field in East Africa and is working with the national park (NP) authorities in three countries to institutionalize the concept of community conservation. The Neighbors as Partners program is operating in Kenya (Tsavo West NP), Tanzania (Tarangire NP, Arusha NP, and the Loliondo area of Serengeti NP), and Uganda (Lake Mburo NP).

The main themes of AWF's Neighbors as Partners program are:

- Identification of people's concerns and problems with wildlife
- Identification of the park managers' problems with adjacent human populations
- Development of the means for dialogue between park managers and the adjacent populations to address issues and resolve problems
- Development of links between park managers and rural development agencies, NGOs, other government departments, research institutions, and other entities that can benefit adjacent communities
- Formation and training of a Community Conservation Service in the park management authority, using a wide range of training techniques based on AWF's experience in the field
- Field assessments of the natural resource base in project areas, both to understand the socioeconomic context and natural resource management problems and to evaluate project effectiveness
- Small-scale development of consumptive and nonconsumptive use of wildlife in selected areas, to reinforce links between the park managers and neighboring communities
- Awareness programs targeted at key government personnel, to demonstrate the need for community wildlife management in regional and national planning

Community conservation seeks to demonstrate and establish wildlife as a natural resource to support rural development. It involves local people in sharing both responsibility for wildlife and the benefits they can derive. AWF is seeking to maintain the integrity of the region's national parks—by facilitating a partnership between park authorities and adjacent communities. AWF's PA:NP program is not a rural development program but may act as a catalyst for such activities.

AWF and "Protected Areas:
Neighbors as Partners" in East Africa

Working with Tanzania National Parks (TANAPA), AWF initiated the PA:NP program in Tanzania in 1987. The long-term objective was to develop a Community Conservation Service within TANAPA, but it was felt appropriate to begin with a pilot project and then move into areas around other parks while establishing and institutionalizing the service within TANAPA. The Loliondo area, east of Serengeti NP, was chosen as the site for the pilot project. The initial objective was to establish communication channels between the park authorities and three Maasai villages adjacent to the park. (It should be noted that these villages were established under a government policy to house rural communities collectively.) At the start of the project, villagers were experiencing considerable cattle rustling by some members of communities west of the park (who illegally used the park as an access and escape route), land in the area was being leased by large business enterprises and grazing areas were thus being reduced, and villagers had no formal land tenure. TANAPA recognized that the Maasai communities and the park authorities both had an interest in allowing land in the area to remain as grazing pasture for livestock and wildlife. TANAPA further recognized that, in the long term, efforts must be made to allow development in the area and that there was an opportunity to demonstrate that wildlife could play a significant role in this development.

A warden was posted to the Lobo area of Serengeti NP (which is adjacent to the Loliondo area). The warden began establishing links with the adjacent villagers and the district authorities. Through village-level meetings, problems were raised, wildlife committees formed, and slowly the main issues emerged. AWF provided assistance in the form of guidance in methods of approaching communities, helping communities to draw up priority issues, and developing systems for reporting back within TANAPA as well as to villagers and district authorities. AWF also worked with the TANAPA warden in developing contacts with other NGO projects in the area and identifying wildlife development options and potential small-scale enterprise partners for these options.

One important outcome of this early work was a request by the villagers to TANAPA for a park ranger post to be sited near the villages. It was felt that the presence of a ranger post would greatly deter cattle rustling activities and also provide antipoaching capability on site to track cattle rustlers if necessary. TANAPA acceded to this request to strengthen the newly emerging partnership by providing a tangible benefit.

AWF assisted in running training workshops for TANAPA wardens to keep them abreast of developments in Loliondo, and to seek assistance from the wardens for ways in which a Community Conservation Service could be established. These workshops focused on the skills needed to be an extension worker, discussions on whether local communities have a role to play in protected area management and how to integrate this role into management planning, how to increase contacts with regional planners, and what kind of training programs are needed to develop a community conservation service.

In 1988, TANAPA put forward a proposal to the Board of Trustees to allow for the establishment of such a Community Conservation Service. This was approved. This then meant that AWF and TANAPA could now plan a longer-term and broader program that would expand into Tarangire and Arusha National Parks and could set up a headquarters unit to coordinate all TANAPA's community conservation activities. The headquarters unit is now established and community wardens are deployed in three parks. A Community Conservation Coordinating Committee has been established by TANAPA to provide a mechanism for advising on the development of the Community Conservation Service and its activities.

Outreach to communities is slowly being expanded through the AWF project and the Serengeti Tourism Education and Extension Project (STEEP). At the same time the community conservation wardens are developing links with relevant local government agencies, with community development projects, and with safari tour operators interested in developing locally based tourism. In 1992 TANAPA introduced Support to Conservation Initiated Projects (SCIPs) for small-scale village-level projects. SCIP funds were set up to reinforce the "problem solving" capacity of community conservation wardens by being applied directly in support of issues identified through TANAPA extension work—but the projects have to be initiated by the communities themselves. The proposed projects have to have a significant local contribution.

The development of AWF's PA:NP project in Kenya was different from the project in Tanzania in that it was not only a process-oriented project but also a problem-oriented one. In this case, a field project was set up in the areas adjacent to Tsavo West National Park, specifically to address the problem of illegal grazing of livestock in the park. Tsavo West shares most of its western border with Maasai group ranches, which in the past were exclusively grazing areas. In recent years portions of these group ranches have been leased or sold as agricultural land, and this loss of grazing land, as well as an accompanying increase in the number of livestock owned by group ranch members, had resulted in the Maasai moving livestock into the national park. This in turn had resulted in serious overgrazing within the park and competition for available grazing between wildlife and livestock. In 1988 AWF appointed a Maasai project officer to work with the Kenyan Wildlife Conservation and Management Department (WCMD) to identify who was grazing cattle in the park and to seek ways to encourage the cattle to be removed. The two-year project mainly comprised a series of boma-level meetings to establish links with the park authorities and to understand the complex issues surrounding grazing rights in the area. The lack of access to markets for cattle and goats is a serious constraint to developing sound ranching practices. Also of importance to the Maasai, and of extreme concern to the wildlife authorities, is the planned subdivision of the group ranches into plots of land with individual title deed. The repercussions for grazing both by livestock and migrating wildlife are grave.

Community Conservation Committees were established on each of the two group ranches in the project area, and a key leader from each community was identified as a Conservation Action Leader (CAL). The committee members and CALs provided a continual link between group-ranch members and the project officer and WCMD staff. The main tangible outputs of the project were the voluntary establishment of a 5-km "border zone" adjacent to the park (by a local chief's order) where only dry-season grazing was allowed and locally based tourism was to be encouraged. The other important output was the voluntary removal of cattle from the park. This had been achieved through many discussions, workshops about grazing resources, developing better links with livestock markets, and introducing potential safari tour operators to the area. (At the end of the project all cattle were outside the park; however, in late 1991 the Maasai negotiated an agreement with the director of the new Kenya Wildlife Service to allow restricted grazing in the park at the end of the dry season, prior to grazing becoming available outside once the rains started. In early 1992 most of the cattle had again left the park.)

During the course of the project, Kenya revamped its wildlife management authority and a parastatal organization, the Kenya Wildlife Service (KWS), was formed. One of the first new policies to be announced by KWS was the intention of sharing revenue from gate receipts with local communities. In July 1990, as a result of its experience in community conservation, AWF was requested by KWS to assist in the development of a policy for wildlife management outside protected areas, and in the development of a community wildlife program. This was done and forms a component of a five-year policy and development program being funded by a consortium of aid agencies. AWF continued its support to KWS by providing assistance in working out mechanisms for sharing revenues acceptable to communities adjacent to the parks. To date mechanisms have been set up to distribute revenue to the Maasai group ranches surrounding Amboseli NP. The process of deciding the most appropriate mechanism took almost one year of negotiation—in a large part because there was no firmly established communication channel between the park authorities and the communities. AWF provided the Tsavo project officer (himself a Maasai) to KWS to act as its representative. Many ideas for mechanisms were mooted, coming from the group ranches, local development NGOs in the area, KWS, and local politicians. The Maasai group ranches have a system that allows consultation with all male members, and this was important in ensuring a consensus was arrived at by each group ranch. Predictably, rivalries between group ranches emerged and local politicians used the opportunity to do considerable electioneering. Eventually KWS suggested how the revenue available might be apportioned in amounts that would be acceptable to all parties. KWS also announced that initially it would solicit and fund community projects that were proposed by the group ranches—in other words, no direct cash payments would be made. This was accepted and the development of project proposals was undertaken with assistance from KWS extension staff.

AWF is assisting in developing a Community Wildlife Service (CWS) within KWS. Inputs are primarily assistance in defining training needs and undertaking training workshops, advising on strategies to deploy personnel with limited extension experience to begin interactions with communities, and general guidance on the CWS' development. AWF is continuing to work in areas adjacent to Tsavo West and to move onto the next stage of ensuring that communities derive tangible benefits from wildlife. Links with private sector tour operators and group ranches are being explored, as well as feasibility studies on game-ranching. In late 1992 AWF (with its partner Development Alternatives Inc.) was awarded a five-year contract to undertake the Conservation of Biodiverse Resource Areas (COBRA) project. This project will build on AWF's work with the KWS Community Wildlife Service and provide an opportunity to consolidate community conservation on a national scale.

AWF's input to community conservation in Uganda began in early 1991 when the Lake Mburo Community Conservation Project was established. Again working with the national park authority, Uganda National Parks (UNP), the project is focusing on small-scale infrastructure support within the park while developing links with communities that live within and adjacent to the park. As in Tsavo, there is a major problem of cattle incursion in the park. The reasons for the incursions are very different and relate back to how the park was originally gazetted in 1982, when more than 300 families with their cattle herds were evicted. No attempt was made to work with the local people and none of the people evicted were compensated in any way or given alternate land on which to settle. The consequence was that the park became surrounded by a highly hostile local population and even became one of the key political points highlighted by the National Resistance Movement in their fight against the Obote government. In 1986 the decision was made to reduce the size of the park by more than half, but to this day cattle and people have remained in the area that is the national park.

Problems such as these need tackling at national policy and regional planning levels as well as on the ground. Extension activities began with the establishment of a community conservation team based in Lake Mburo. The team has worked closely with local political and administrative leadership as well as with communities around the park. Notable achievements have been the development of a park management plan that included comprehensive community consultations—the first time this has been done in Uganda—and the proposed establishment of an advisory committee to provide the park and communities a structure for ongoing planning. SCIP funds, similar to the system established by TANAPA, were also set up.

Constraints and Responses

The above summaries of the projects in Kenya, Tanzania, and Uganda briefly describe the activities and approaches undertaken. What is perhaps more interesting in the context of how model projects might be replicable in other areas, is the constraints such projects face and the ways that AWF has tried to tackle these. Experiences in Tanzania, for example, have been useful when designing projects in Kenya, and vice versa. The major actions found for overcoming the constraints of community conservation projects are discussed below.

1.   *Involve government.* In any project where activities become multisectoral and beyond the confines of an agency's traditional sphere of expertise, it is important that there is significant government involvement. Identifying which sector of government to work with is an integral part of community conservation work, but initially the protected area management authority is the most obvious partner. AWF therefore sought to institutionalize the concept within the park management body from the start. By doing this, a long-term perspective could be brought to the projects, which is essential when embarking on community projects of this nature.

2.   *Elicit local participation.* "Local participation" has become somewhat of a buzzword in community conservation work. Everyone emphasizes the need for communities to become involved in projects, from planning to execution. What is perhaps not appreciated by park management authorities,

who suddenly see an urgent need to seek partnership with local communities, is the time needed to build up links with those communities. Communities adjacent to national parks are not homogeneous and require time to build up a consensus among themselves. Any community has issues of importance and key actors, and the ties and pulls by the different actors with different agendas will affect local participation in conservation.

At this level park authorities need to recognize that communities can be functionally defined in several ways, such as representative structure, common interest, land use, those who pay the highest cost in terms of the park's existence or pose the greatest threats. This will help provide priorities and target communities as natural resource management institutions.

Many communities are suspicious of park authorities because of past negative interactions. Exceptionally good extension work is usually needed to begin to develop communication channels, to begin to enable communities to make informed choices about their use of natural resources, and then to encourage them to work with the wildlife authorities in promoting sustainable use of the wildlife resources. The wildlife authorities can also act as a catalyst to spur activities in other aspects of natural resource management, such as soil erosion control and reforestation projects. AWF has found that the process of building links with communities has taken a long time but, provided consistent sensitive inputs are made, momentum can be maintained.

3.   *Catalyze rural development.* One of the most important factors to be understood in community conservation is the need for the wildlife management authority to be a catalyst and facilitator and stimulate community involvement in development and conservation activities relevant to issues of the area. There is a danger that wildlife authorities, as they develop community conservation projects, may go beyond the point where they have a legitimate interest or comparative advantage and move into trying to become a rural development agency, rather than just promoting the idea that wildlife has a role in development. In the end it is unlikely that wildlife authorities can, or should, play the role of a development agency around the borders of all their protected areas. In addition, naive approaches to rural development may mean that many past mistakes are perpetuated by well-meaning people. It is important to structure community conservation projects with a view to linking up with existing networks, or to forming partnerships with other agencies skilled in community development techniques, and focus on promoting how the wildlife resource can be a positive factor in an area's development. This is obviously made easier if community development agencies are already established in an area and have done much of the work in understanding a community's needs. This has been the approach of the CAMPFIRE program and it is the direction that AWF's PA:NP program is moving toward now that

the concept of community conservation is established within the park management authorities in East Africa.

4.   *Choose target groups carefully.* Most community conservation projects focus only on the communities adjacent to the protected areas in the belief that these people are the resource users. However in many places the pressures on natural ecosystems come from many directions and are often related to factors over which rural people have no control—for example, laws, policies, and economic factors. As part of its PA:NP program AWF has deliberately taken a vertically integrated approach—grassroots activities combined with interaction at each country's regional planning levels as well as advice on policy at national level. In Tanzania an AWF and WWF project on Planning and Assessment for Wildlife Management is based within the Wildlife Division and is seeking to increase the contribution wildlife makes to the Tanzanian national economy as part of the Economic Recovery Program. This project will hopefully develop policy to create an enabling environment for village-utilization schemes and will thus complement and back up the work being undertaken on the ground by TANAPA. Similarly in Kenya, when the opportunity arose, AWF assisted in developing policy that will allow revenue sharing to reach appropriate targets as well as encourage wildlife to be a more positive factor in rural Kenyans' lives through limited-use rights on private lands.

5.   *Define mandate, and increase resources, equipment, and trained personnel.* In many countries protected area management authorities lack resources, equipment, and trained personnel and the establishment of community conservation programs place additional strains on small budgets. Traditional wildlife management training has not equipped nor even sensitized park managers to recognize the needs of local communities. In addition, in many places, it is not always clear whether park authorities have jurisdiction to operate outside their national parks. There is a need to sort out at national policy level who has broad responsibility for what so that the mixture of parks, reserves, biosphere reserves, and so on mean something to the people.

6.   *Seek self-sufficiency.* Many community conservation projects are designed with notions of self-sufficiency, but this has yet to become widespread even at project sites where the potential for income-generating projects linked to wildlife is high. AWF has therefore placed great emphasis on promoting the idea of community conservation as a necessary component of modern day park management and the need to recognize the associated costs. This has not been the emphasis of other similar projects on the continent.

At the early stages of such development it is difficult to calculate budgetary and manpower implications, so AWF has provided financial assistance and technical expertise to begin the institutionalizing—the pivotal component of the PA:NP program. Part of the need for trained personnel has been addressed by developing a new course at the College for African Wildlife Management at Mweka in Tanzania. The "Man and Wildlife" course was taught for the first time in 1989 and, although only an introduction to the concept of community conservation, has already stimulated graduates to ask to be posted to do community work. Such approaches need to be expanded into other disciplines, such as forestry.

If there will always be an associated cost for community conservation for a wildlife management authority then each country needs to define acceptable limits to that cost and work toward developing a program within these limits. It will need several more years of projects on the ground before estimates can be made. It may be that donors will continue to support initiatives that bring the wildlife resource into development activities, but there is a need to educate donors that such support will need to be long-term if it is to be meaningful and have a chance to succeed. This is very important as most donors commit funds for short to medium periods (up to 5 years) and require outputs to be listed in the project design that may be inappropriate and unobtainable within the funding period. The design of most community conservation work would benefit from long-term funding (10–20 years), and current projects should work to identify relevant methods of monitoring and evaluating progress to justify longer term financial commitment.

7.   *Consider alternative methods of resource utilization.* For community conservation projects to succeed in the long term, solutions to problems of overexploitation of natural resources will also need to be long term. While the AWF project in Tsavo succeeded in getting the livestock out of the park for a short period, the present stocking levels on the group ranches has meant that cattle returned to the park in the extended dry season of late 1991. Clarification of the responsibility of livestock owners to undertake sound management needs to be developed. At Lake Mburo, resolving issues of land tenure outside the park, availability and access to water, and resettlement of communities in the park are essential for the long-term survival of the park. For example, if adjacent communities fail to secure title for their land, no preventive measure to address the increasing problem of soil erosion will be undertaken, the quality of the grass layer outside the park will deteriorate, and incursions by livestock will increase in duration. Wildlife management, both inside and outside protected areas, and land-tenure systems are inextricably linked; wildlife authorities must address this and build high-level and local-level links with bodies responsible for land use.

Conclusion

The establishment of community conservation services within Tanzania National Parks and the Kenya Wildlife Service is already having, and will continue to have, a large impact as it has facilitated structured partnership links between the two park authorities and the communities living adjacent to their parks. Effective and trusted communication channels need to be in place long before efforts can be begun to promote community organization and start introducing ideas of economic benefits. This has been achieved in the project areas in both Tanzania and Kenya.

With channels in place in several parks and methodologies defined for moving to new areas, TANAPA and KWS have now initiated revenue sharing activities, using their community conservation staff to work out appropriate mechanisms. The next stage will be assisting communities to realize economic opportunities, and funds are already available for Kenya to allow these options to be developed. At this stage the impact of community conservation will be more discernible.

AWF appreciates the need to monitor and evaluate initiatives and to try and measure the impact the PA:NP program is leaving. But one must understand that the more people are involved in decision making—the more they participate—the less one can monitor and evaluate in the traditional sense of impact and cost benefits.

In East Africa, a number of promising trends can be seen.

- More structured partnerships are being built between parks and people.

- Training within the parks' services and at a regional wildlife management training institution now includes community conservation.

- National wildlife authorities are starting to make a commitment to revenue sharing.

- Governments at national level have a greater awareness (and so responsibility) for environmental conservation and sustainable management, and they show a commitment to conservation in partnership with local communities.

These are important trends that, together with a real participatory involvement of local people and their knowledge and skills, will increasingly show tangible results. The challenge now is to consolidate the experience and help devise the means to institutionalize this on a local, national, and indeed international basis.

Index

Island Press Board of Directors